PLANT LIFE CYCLES

PLANT LIFE CYCLES

THOMAS R. MERTENS
FORREST F. STEVENSON
Ball State University
Muncie, Indiana

John Wiley & Sons, Inc.
New York • London • Sydney • Toronto

Editors: Judy Wilson and Irene Brownstone
Production Manager: Ken Burke
Editorial Supervisor: Winn Kalmon
Artist: Forrest Stevenson
Composition and Make-up: Wendy Welsh

Library of Congress Cataloging in Publication Data

Mertens, Thomas Robert, 1930-
Plant life cycles.

(Wiley self-teaching guides)
Bibliography: p. xv
Includes index.
1. Plants--Reproduction--Programmed instruction.
2. Botany--Programmed instruction. I. Stevenson, Forrest Frederick, 1916- joint author. II. Title.
(QK825.M47) 581.1'6'077 74-19373
ISBN 0-471-59629-9

Fred Ag

To the Reader

Plants are different from animals! To most of us this statement is so obvious that we would not think of challenging it. At the same time most of us, unless we are professional biologists, would be hard pressed to specify the basic ways in which plants differ from animals. And there are many similarities. So the problem becomes one of clarifying the differences and the similarities so that the nonprofessional can understand them and appreciate their significance.

Most of us know a great deal more about the biology of animals than we do about the biology of plants. We are familiar with the reproductive process of our pets and other domestic animals, and with the human reproductive process. We may even be quite surprised to find that sexual reproduction occurs in most plants. We may also expect that when it does occur the mechanisms will be identical to those in the animals with which we are familiar.

Plants, of course, are fascinating in their own right, as any backyard gardener, house plant enthusiast, or spring wild flower photographer would tell you. All of these people—and the student of biology or botany as well—will find that the study of plant life cycles is a fascinating subject. Some of the simpler plants reproduce without sex, but most plants have at least some stage in their life cycle where they reproduce sexually. How do plants differ from animals with respect to sexual reproduction? How did plants come to evolve the unique life cycles that are so characteristic of them? How did certain special reproductive structures, such as seeds, come into existence? These are some of the questions that this book will answer.

Experience in teaching about plant reproduction and life cycles suggests that, although plants are different from animals, the learner often tries to transfer understanding of animal reproduction to that of plants. This often results in a misunderstanding of plant reproduction.

Study of this book should set the record straight. We have tried in the first chapter to capitalize on your understanding of human and animal reproduction to show you what similarities and differences exist when plant and animal reproduction are compared. We also consider the evolutionary significance of sexual reproduction to living organisms and introduce the underlying pattern of sexual reproduction in plants—alternation of generations. This pattern unites the remaining chapters as we examine how more advanced plants reproduce. Beginning with liverworts and mosses and progressing to Selaginella, cycads, pines, and finally to the flowering plants, we see that all have life cycles of a gametophyte generation alternating with a sporophyte generation.

When you have finished the book, you should have a clear understanding of plant reproduction and life cycles.

September, 1974

T. R. M.
F. F. S.

How to Use This Book

This book is designed so that you can learn about plant life cycles enjoyably and thoroughly at your own pace. The material is presented in sections called frames, each containing questions and answers that let you test your understanding of the new material. As you come to a question, pause and answer in the space provided. Your answer may be a true-false or multiple-choice selection, a word or short phrase to fill in a blank, an explanation, or a simple diagram; it may involve either reasoning or remembering. Check your answer with the answer given beneath dashes that separate each question from its answer. For maximum learning, try not to look at the printed answer until you have first given your own. (You may want to cover the answers with a separate sheet of paper or a 3 x 5 card while you are working on the question.)

You will be able to answer practically all of the questions as you come to them because the questions progress from simple to complex and each frame builds upon information in the previous frames. This learning program is not a workbook and it is not a textbook. Unlike a workbook, this program is self-contained and provides answers for all the questions it asks. Unlike a textbook the program helps you to learn and, through frequent problems and review items, helps you to be confident that you can use the material it presents.

Work at your own pace. You may be able to move rapidly through some sections but only slowly and thoughtfully through others. Some learners may complete the program in four hours, others in sixteen, a typical figure probably being eight or ten hours—probably less time than you would spend learning the same material from textbook reading assignments, class lectures and discussions, and studying for a test.

Do only one chapter at a time, with a break (even a day) in between. Because of the continual review in answering questions, you will not forget one chapter if you wait awhile before beginning the next.

Review items occur throughout the program, and each chapter concludes with a Self-Test. If you are in a biology or botany course, you will find all of these sections helpful in preparing for a classroom test. A Final Test is provided at the end of the book, so that you can evaluate your learning.

Contents

Behavioral Objectives and Field Testing

To assist the reader in learning the content of this book, we have prepared a precise set of behavioral or instructional objectives. These objectives indicate the kinds of behaviors that the student should be able to demonstrate once the content of the book has been mastered. Upon completion of this book the student will be able to:

- Compare and contrast sexual and asexual reproduction, citing the unique features of each, and describing the evolutionary advantages of sexual over asexual reproduction.
- Describe the unique role of meiosis in the life cycle of a sexually reproducing organism, contrasting the consequences of meiosis and fertilization on chromosome number.
- Compare and contrast heterogamy and isogamy, citing examples of organisms having each of these types of sexual reproduction.
- Outline alternation of generations as it occurs in the life cycles of moss, liverwort, fern, *Selaginella*, cycad, and pine, using both words and diagrams.
- Describe the significance of heterospory in producing specialization in the life cycles of *Selaginella*, gymnosperms, and angiosperms.
- List four major adaptations which a plant must possess to be a seed-producing plant.
- Cite the evolutionary significance of seed production for the survival of terrestrial plants.
- Distinguish between the origins of the nutritive tissue surrounding the embryos in the seeds of gymnosperm and angiosperm plants.
- Outline alternation of generations as it occurs in an angiosperm (flowering plant), noting the modifications and specializations that occur in flowering plants as compared to moss, liverwort, fern, *Selaginella*, cycad, and pine.

- Apply knowledge of the life cycle of an angiosperm to determination of chromosome number in various cells in the gametophyte and sporophyte generations, given the chromosome number in a cell of one of these stages.

- Compare and contrast the gametophyte and sporophyte generations of moss, liverwort, fern, Selaginella, cycad, and flowering plant with respect to chromosome number, method of reproduction, and relative size or conspicuousness.

- Apply knowledge of alternation of generations to the interpretation of the life cycle of an "unknown" or previously undescribed plant, given a minimal amount of data about that plant, its parts, reproductive behavior, and life cycle. (For example, given certain minimal information about the life cycle of a fern, horsetail, or cycad, for instance, the student should be able to fit this life cycle into the scheme of alternation of generations, describing what parts are haploid and diploid, indicating where meiosis occurs in the cycle, where mitosis and fertilization take place, and so on).

- Use the vocabulary of plant reproduction summarized below to outline the life cycles of representative plant types.

Chapter 1

alternation of generations
analogous
asexual reproduction
chromosome
chromosome number
DNA/deoxyribonucleic acid
diploid (2n)
egg
fertilization
flagellum (pl., flagella)
gamete
gametophyte
gametophyte generation
gene
haploid (n)
heterogametes/heterogamy
homologous pairs of chromosomes
isogametes/isogamy
meiosis
mitosis
morphological
nucleus (pl., nuclei)
Oedogonium
physiological
propagated
sexual reproduction
sperm
spore
sporophyte
sporophyte generation
Ulothrix
Ulva
zygote

Chapter 2

aquatic
capsule
club moss
cone
conspicuous/inconspicuous generation
dependent
fern

Chapter 2
(continued)

liverwort
Marchantia
moss
seta
sporangium (pl., sporangia)
sporophyll
strobilus (pl., strobili)
terrestrial
vascular tissue

Chapter 3

heterospory/heterosporous
homospory/homosporous
megagametophyte (female gametophyte)
megasporangium (pl., megasporangia)
megaspores
megasporophyll
microgametophyte (male gametophyte)
microsporangium (pl., microsporangia)
microspores
microsporophyll
Selaginella

Chapter 4

dormancy
embryo
functional megaspore
integuments
megaspore wall
micropyle
nucellus
ovule
pollen grain
pollen tube
pollination
seed
seed coat

Chapter 5

conifer
cover scale
cycad
fir
functional megaspore
gymnosperm
megaspore mother cell
ovulate strobilus (cone)
ovuliferous scale
pine
primitive vs. advanced characteristics
spruce
staminate strobilus (cone)
woody scale

Chapter 6

angiosperm
anther
bisporangiate strobilus
carpel
double fertilization
embryo sac
endosperm nucleus
essential/nonessential flower parts
filament
flower
fusion nucleus

Chapter 6
(continued)

ovary
petal
pistil
polar nuclei
pollen sac
receptacle
sepal
simple/compound pistil
stamen
stigma
style
tricarpellate
triploid
whorl (layer)

Plant Life Cycles was field-tested with 45 beginning plant science students in the Department of Biology at Ball State University, Muncie, Indiana. On the first instructional day of the spring quarter, 1974, each of the students was given a 30-item pretest. The mean score on this test was determined to be 8.4. After a two-week interval for study of Plant Life Cycles an identical post-test was administered to the students. The mean score on the post-test was calculated to be 21.8. A gain score may be calculated using the following formula recommended by Professor McGuigan of Hollins College, Virginia:

$$\% \text{ Gain} = \frac{\text{Post-Test Score Minus Pretest Score}}{\text{100\% Minus Pretest Score}}$$

Thus, using the scores indicated above,

$$\% \text{ Gain} = \frac{21.8 - 8.4}{30.0 - 8.4} \times 100 = \frac{13.4}{21.6} \times 100 = 62\%$$

The pre- and post-tests were modified and combined to form the final test in this book. Plant Life Cycles was revised according to students' testing experience, as well as subject and programming reviewers' comments, to clarify where necessary before publication.

The authors and publisher are very interested in data on the use of this book in other settings. Please send any pre- and post-test scores on Plant Life Cycles, along with any other comments or suggestions, to the Self-Teaching Guides Editor, John Wiley & Sons, 605 Third Avenue, New York, New York 10016.

Reference Chart for Selected Biology and Botany Textbooks

You may find the information in the table on page xvii useful if you wish to pursue certain topics in greater depth or if you are a student in a course using one of the books listed below. The table correlates chapters in this book with material on the same subjects in a number of the newer basic biology and botany textbooks. The numbers listed in the table correspond to chapter (and part) numbers in the textbooks.

Baker, J.J.W. and G.E. Allen, The Study of Biology, 2nd ed. (Reading, Massachusetts: Addison-Wesley, 1971).

Bold, H.C., Morphology of Plants, 3rd ed. (New York: Harper and Row, 1973).

CRM Books Editorial Staff, Biology Today (Del Mar, California: CRM Books, 1972).

Cronquist, A., Basic Botany, 3rd ed. (New York: Harper and Row, 1973).

Curtis, H., Invitation to Biology (New York: Worth, 1972).

Jensen, W.A. and F.B. Salisbury, Botany: An Ecological Approach (Belmont, California: Wadsworth, 1972).

Keeton, W.T., Biological Science, 3rd ed. (New York: Norton, 1972).

Keeton, W.T., Elements of Biological Science, 2nd ed. (New York: Norton, 1973).

Neushul, M., Botany (Santa Barbara, California: Hamilton, 1974).

Raven, P.H. and H. Curtis, Biology of Plants (New York: Worth, 1970).

Stephens, G.C. and B.B. North, Biology (New York: Wiley, 1974).

Weier, T.E., C.R. Stocking and M. G. Barbour, Botany—An Introduction To Plant Biology, 5th ed. (New York: Wiley, 1974).

Wilson, C. L., W. E. Loomis, and T. A. Steeves, Botany, 5th ed. (New York: Holt, Rinehart and Winston, 1971).

Chapter in this book	Baker & Allen	Bold	CRM Books Staff	Cronquist	Curtis	Jenson & Salisbury	Keeton (Science)	Keeton (Elements)	Neushul	Raven & Curtis	Stephens & North	Weier, Stocking, & Barbour	Wilson, Loomis, & Steeves
1. Sexual and Asexual Reproduction	15, 25	2	13, 16, 37	10, 19, 20, 21	3(1)	9	13, 21	13, 21	7, 8, 9, 10, 11	5(2), 6(4)	7, 12	22	11, 19
2. Alternation of Generations	25	16, 17	37	12, 15, 16	4	23	21	21	12	6(5)	7, 12	22, 26	23, 24
3. Heterospory	25	19	37	16		25	21	21	14	6(6), 6(7)	12	27	26
4. Formation of Seeds	25	25	37	17, 25		27, 28	21	21	15	6(6)	12	15, 16, 28, 29	12, 27, 28
5. Gymnosperms	25	25, 27, 29	37	17		27	21	21	16	6(7)	7, 12	28	27
6. Angiosperms	25	30, 31	37	25	5(1), 5(2)	28	21	21	17	6(7), 6(8)	7, 12	15, 16, 29	12, 28

PLANT LIFE CYCLES

CHAPTER ONE

Sexual and Asexual Reproduction

1. Sexual reproduction in animals is widely understood among non-biologists. All of us recognize the sexual nature of human reproduction and we realize that similar mechanisms occur in our domestic pets, farm animals, and in wild animals. We know that in these animals a new individual arises when an egg cell from the mother unites with a sperm cell produced by the father. The union of these special cells is characteristic of sexual reproduction. Collectively, these cells are called gametes; the egg and the sperm are specific examples of gametes.

   Many people would be greatly surprised to learn that some animals can reproduce asexually—without sex. In certain breeds of turkeys, for example, offspring develop from unfertilized eggs. The male honeybee—the drone—is typically reproduced in this fashion; it has only one parent, its mother, since the drone bee develops from a female gamete (egg) that has not been fertilized by a male gamete (sperm).

   a. From the discussion in the preceding paragraphs we can conclude that in sexual reproduction, how many parents are typically involved? ____________________

   b. The general term used to refer to sex cells that unite during sexual reproduction is ____________________.

   c. Sexual reproduction typically involves the union (fertilization) of a/an ____________________ cell with a/an ____________________ cell.

- - - - - - - - - - - - - - - - - - -

a. two; b. gametes; c. egg, sperm (either order)

2. The non-biologist is often greatly surprised to learn that plants also reproduce sexually. The gardener or the house plant enthusiast knows that geraniums can be propagated or reproduced by rooting "cuttings" from an old plant in moist sand, or that a new African violet can be started by rooting a leaf in a glass of water. Obviously in such examples of plant reproduction, there is no union of an egg cell with a sperm cell. How many parents are involved when plants are

   reproduced from cuttings? ____________________

   \- - - - - - - - - - - - - - - - - - -

   one

3. What these people may not realize is that the <u>seed</u>, probably the most universally recognized mechanism of plant reproduction, is formed after the fertilization of a female gamete or egg by a male gamete or sperm. Seeds, then, are formed as a result of (sexual/asexual)

   ____________________ reproduction, since they are produced when

   certain cells called ____________________ unite.

   \- - - - - - - - - - - - - - - - - - -

   sexual; gametes (or sperm and egg)

4. In this book we will be looking in detail at various mechanisms of reproduction in plants and at how these mechanisms affect the entire life cycle of plants. In particular, we will stress <u>sexual</u> reproduction in plants.

   At this point in your study, you know that sexual reproduction

   occurs when sex cells called ____________________ unite. The union

   of these cells is called (a cutting/fertilization) ____________________.
   When a plant or animal is reproduced by a mechanism that does not involve the union of sex cells, we say that the reproduction has oc-

   curred ____________________ (literally, "without sex").

   \- - - - - - - - - - - - - - - - - - -

   gametes (or sperm and egg); fertilization; asexually

5. The development of three starfish from one starfish that has been fragmented into three parts or the propagation of an African violet by rooting a leaf are both examples of what kind of reproduction?

____

- - - - - - - - - - - - - - - - - - -

asexual reproduction

6. To summarize, in asexual reproduction the offspring or new individual is derived directly from its parent with no union of sex cells called gametes. The offspring produced by asexual reproduction is essentially a part of the parent that has become separated from the parent and now exists as a new individual.

   Note, however, that since the offspring which is produced asexually has been derived directly from its one parent, it has received all of its heredity or inherited (genetic) traits from that one parent. Thus, we could expect the asexually produced offspring to be (different from/similar to/identical to) ____________________ its parent.

- - - - - - - - - - - - - - - - - - -

identical to

7. The gardener who has an especially desirable variety of plant with large, showy flowers or the horticulturist desiring to propagate a fruit tree with an especially tasty fruit can be certain that with asexual reproduction the variety of plant can be reproduced in a form identical to the desirable parent variety. How can they be sure?

____

____

- - - - - - - - - - - - - - - - - - -

They can be sure because an organism that reproduces asexually inherits all its traits from its one parent.

8. Similarly, asexual reproduction offers certain advantages to the plant species itself. If the species is well adapted to its environment and flourishes in that environment, asexual reproduction will result in offspring that are identical to the parent and thus (unfit for/adapted to)

____________________ the environment in which the plant species is found.

\- - - - - - - - - - - - - - - - - - -

adapted to

9. By the same token, however, a plant species that reproduces asexually and is adapted, let's say to a tropical environment, would find it difficult to become established in a temperate environment. An asexually reproducing plant adapted to a tropical environment would produce offspring that are (different from/similar to/identical to) ____________________ itself and thus adapted only to the (aquatic/desert/tropical/temperate) ____________________ environment.

\- - - - - - - - - - - - - - - - - - -

identical to; tropical

10. A plant species that reproduces asexually may be very well adapted to certain environmental conditions, but would not be suited to survival if the environment were to change. As a consequence, we must conclude that asexually reproducing organisms are, in the long run, at an evolutionary (advantage/disadvantage) ____________________.

    Why? ____

    ____

\- - - - - - - - - - - - - - - - - - -

disadvantage
because they cannot adapt to changing environmental conditions

11. One of the chief advantages to a species that reproduces sexually is that sexual reproduction results in new combinations of traits by combining the characters of the parents in new ways. For example, corn is a sexually reproducing plant which possesses a great deal of genetic variability. If a tall corn plant with crinkly leaves is crossed with a dwarf plant having normal leaves, entirely <u>new</u> combinations of these traits may be found in the offspring. Thus, one might expect to find <u>tall</u> plants with <u>normal</u> leaves as one new type.

What other new type might be produced from this cross?

___

- - - - - - - - - - - - - - - - - - -

dwarf plants with crinkly leaves

12. Herein lies the reason why sexual reproduction is so widespread among both plants and animals: sexual reproduction, with its typical mixing of hereditary or genetic traits from two parents, insures a great deal of (similarity/variability) ___ among the offspring.

- - - - - - - - - - - - - - - - - - -

variability

13. A variety of biological investigations show that plants and animals do not change or adapt in order to adjust to changing environmental conditions. Rather those organisms having gene combinations that permit them to survive in a changed environment are the ones selected for survival. Proof of the variety occurring in sexually reproducing organisms can be seen in the human species, where each individual (other than identical twins) is unique and different from every other individual. The advantage that sexual reproduction confers on a species is the variation that (permits/prevents) ___ adaptation to different environmental conditions.

- - - - - - - - - - - - - - - - - - -

permits

14. Over a long period of time, great environmental changes may occur. Which would have a better chance of survival—an asexually reproducing species or a sexually reproducing species? ___

    Explain. ___

    ___

    ___

- - - - - - - - - - - - - - - - - - -

a sexually reproducing species
Your answer should have included these key points: that a sexually reproducing species has a better chance of survival because it can adapt to environmental changes, but an asexually reproducing species cannot so readily adapt and may be eliminated.

15. To summarize some of the ideas thus far presented, complete the following matching exercise.

______ a. gamete		1. fertilization
______ b. insures offspring identical to parent		2. without sex
______ c. union of gametes		3. sexual reproduction
______ d. sperm		4. male gamete
______ e. asexual		5. asexual reproduction
		6. adaptation
		7. general term referring to cells which unite in reproduction

- - - - - - - - - - - - - - - - - - -

a. 7; b. 5; c. 1; d. 4; e. 2

16. Perhaps this is an appropriate place to introduce another aspect of sexual reproduction. We have noted cases where sex cells—gametes—are clearly differentiated into a male gamete and a female gamete. In these cases it is proper to use the general term heterogametes (literally, "different gametes"), since these cells are clearly differentiated into male sex cells or ____________________ and female sex cells or ____________________.

- - - - - - - - - - - - - - - - - - -

sperms; eggs

17. In many simple organisms, however, gametes are produced that cannot be differentiated by microscopic examination into sperm and egg. These gametes are technically called isogametes (literally, "same gametes") because they appear to be (different/similar/identical) ____________________ under the microscope.

- - - - - - - - - - - - - - - - - - -

identical

18. Which characterizes human sexual reproduction—isogamy or heterogamy? ____________________

- - - - - - - - - - - - - - - - - - -

heterogamy, since human gametes are clearly differentiated into sperm and egg.

19. The plant body of the freshwater alga called Ulothrix (diagram below) consists of an unbranched string of cells called a filament. Special cells in the filament produce from 8 to 64 isogametes—small cells identical in appearance and possessing two whiplike flagella, which enable the gametes to swim. Thus, when Ulothrix reproduces sexually, it does so by the union of ____________________.

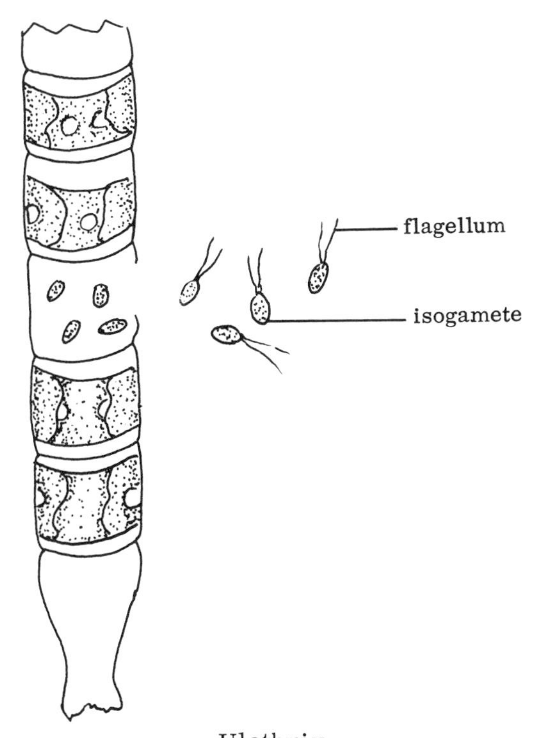

Ulothrix

- - - - - - - - - - - - - - - - - - -

isogametes or sex cells that are identical in appearance

20. Although these isogametes are morphologically identical (they appear identical), they are physiologically or functionally different. Gametes from one Ulothrix filament always unite with gametes from a different Ulothrix filament. We cannot properly speak of Ulothrix gametes as sperm and egg, because they are morphologically identical. We could signify the functional difference that does exist by designating them as + and - gametes.

    Thus, it would seem reasonable to conclude which of the following? (Check those that are correct).

    ______ a. Plus gametes will unite only with plus gametes and minus with minus.

    ______ b. Plus gametes will unite only with minus gametes.

    ______ c. There is true sexual reproduction in Ulothrix because there is a union of sex cells.

    ______ d. There is no true sexual reproduction in Ulothrix, since sperms and eggs do not occur in this alga.

- - - - - - - - - - - - - - - - - -

b and c are correct

21. Another freshwater filamentous alga, Oedogonium, presents a classical example of heterogamy. This means that this alga should produce gametes that are clearly differentiated into ______________ and ______________ when viewed microscopically.

- - - - - - - - - - - - - - - - - -

sperm; egg (either order)

22. Are the gametes of Oedogonium different morphologically? ________ ________ Are they different physiologically? ______________

- - - - - - - - - - - - - - - - - -

yes, they look different;
yes, the sperm and egg function differently

23. In Oedogonium, as in other organisms reproducing by heterogamy, small, mobile sperm cells and relatively large, non-mobile egg cells are produced, with the sperm swimming to where the egg is produced and uniting with it in the process called ____________________.

- - - - - - - - - - - - - - - - - - -

fertilization

24. Gametes that are similar in size, shape, and locomotion are called ____________________, while those that are clearly differentiated into a relatively large immobile egg and a small mobile sperm may be collectively called ____________________.

- - - - - - - - - - - - - - - - - - -

isogametes; heterogametes

25. The alga Oedogonium reproduces by heterogametes, while the filamentous alga Ulothrix reproduces by ____________________.

- - - - - - - - - - - - - - - - - - -

isogametes (or isogamy).

26. Let's turn our attention to another aspect of sexual reproduction and consider chromosomes and chromosome number as they relate to gametes and fertilization. You may recall from other studies that chromosomes are darkly staining bodies found in the nucleus of the cell. You may also know that chromosomes are important in reproduction because they bear the hereditary material DNA (deoxyribonucleic acid) from parents ot offspring. Shown on the next page is a diagram of a root tip cell from a corn plant as it might appear when viewed with a microscope at the time it is about to undergo cell division. At this time the chromosomes can be stained so that they are readily visible. How many chromosomes can you count in the cell? ____________________

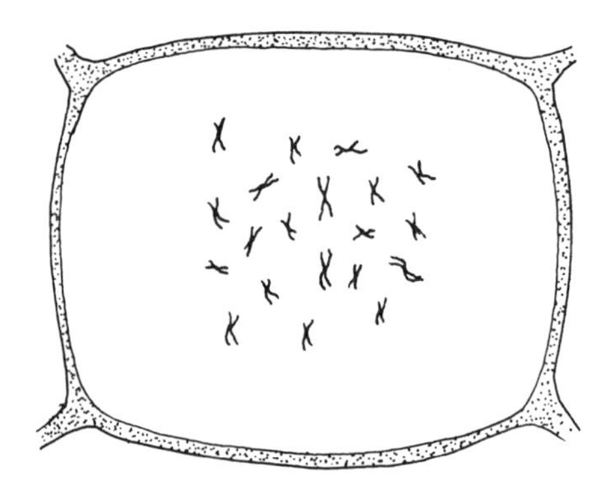

- - - - - - - - - - - - - - - - - - -

There are 20 chromosomes in the cell.

27. The number of chromosomes found in the cell nuclei of a certain kind of plant can be counted. Corn plants typically have 20 chromosomes in all of the cells of the plant body, while tomato plants always have 24, and pea plants always have 14. Similarly, humans have 46 and dogs have 78 chromosomes in the nuclei of their body cells. Thus, we may conclude that for different species of organisms chromosome number is (constant/variable) ______________________. However, within a single species chromosome number is (constant/variable) ______________________.

- - - - - - - - - - - - - - - - - - -

variable; constant

28. The cell nucleus contains a number of countable, darkly staining structures which are important because they bear the hereditary material from cell to cell and from parents to offspring. These structures are called ______________________.

- - - - - - - - - - - - - - - - - - -

chromosomes

29. Now consider the following problem. Suppose two dogs were mated. If the female gamete (egg) and the male gamete (sperm) each carried the same number of chromosomes (78) as the body cells of the dog, how many chromosomes would be present in the cells of the offspring?

____

\- - - - - - - - - - - - - - - - - -

78 in egg + 78 in sperm = 156 in cells of offspring

30. Obviously, if this mechanism were to prevail, the chromosome number would double in each generation of sexual reproduction. Since human origins date back many generations to our prehistoric ancestors, you can imagine how many chromosomes would be present in the cell nuclei of modern humans! Obviously, this just won't work! There must be some mechanism to keep chromosome number constant for a species. This mechanism is a special kind of nuclear division called meiosis. The net effect of the meiotic division of a nucleus is the production of daughter nuclei that are haploid—that have half the typical number of chromosomes. In animals meiosis occurs when gametes are formed. As you will learn later meiosis occurs at a different time in the life cycles of most plants. Nonetheless, gametes in both plants and animals are haploid in chromosome number. Thus, if human beings have a typical chromosome number of 46, we can conclude that human gametes produced by meiosis would contain how many chromosomes? ______________________

\- - - - - - - - - - - - - - - - - -

23

31. The chromosome numbers for the various organisms discussed in frame 27 are called the diploid numbers. These are double the number of chromosomes in the gametes. The haploid number is frequently designated as the n number, and the diploid number referred to as 2n.

If the plant body cells of the common garden onion each contain 16 chromosomes, what is the diploid (2n) chromosome number for this plant? ______________________ What is the haploid (n) number?

____

\- - - - - - - - - - - - - - - - - -

16; 8

32. The human sperm contains ____________________ chromosomes, the egg contains ____________________ chromosomes, and the fertilized egg or zygote must, therefore, contain ____________________ chromosomes. We say that the gametes are (haploid/diploid) ____________________, while the zygote is (haploid/diploid) ____________________.

- - - - - - - - - - - - - - - - - - -

23; 23; 46; haploid; diploid

33. Meiosis and fertilization are thus opposing processes. (Meiosis/Fertilization) ____________________ reduces the chromosome number, while (meiosis/fertilization) ____________________ restores the diploid number of chromosomes in the fertilized egg.

- - - - - - - - - - - - - - - - - - -

Meiosis; fertilization

34. In summary, meiosis is a universally occurring process in sexually reproducing organisms. Meiosis coupled with ____________________, which restores the diploid chromosome number, maintains chromosome number at a (constant/variable) ____________________, generation after generation.

- - - - - - - - - - - - - - - - - - -

fertilization; constant

35. To illustrate this summary of the relationship between meiosis and fertilization, study the following diagram of the life cycle of a mouse and answer the questions asked.

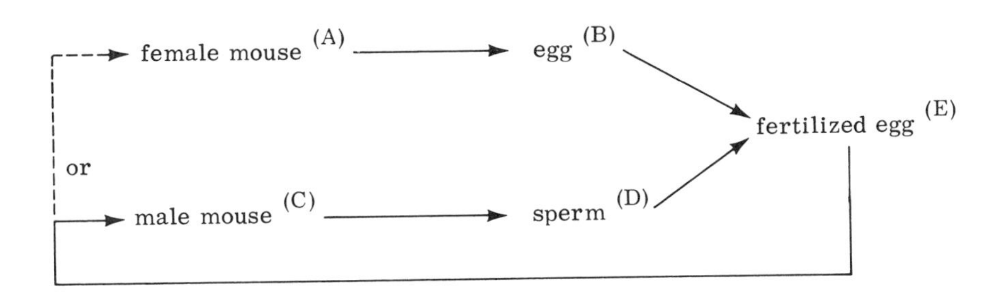

Between what stages (indicated by letters) in this diagram does the the process of meiosis occur? ________________ Between what letters in this diagram does the process of fertilization occur? ________________ If the diploid chromosome number in the mouse is 40, what is the chromosome number of each of the following cells—egg? ________________ sperm? ________________ fertilized egg? ________________

- - - - - - - - - - - - - - - - - - -

Meiosis occurs in the female between A and B and in the male between C and D.
Fertilization occurs when B and D unite to form E.
20; 20; 40

36. We have indicated that body cells are diploid or (4$\underline{n}$/3$\underline{n}$/2$\underline{n}$/$\underline{n}$) ______. Cells produced as a result of meiosis have the "half-number" of chromosomes and are said to be ________________. Such cells have the (4$\underline{n}$/3$\underline{n}$/2$\underline{n}$/$\underline{n}$) ______ number of chromosomes.

- - - - - - - - - - - - - - - - - - -

2$\underline{n}$; haploid; $\underline{n}$

37. Chromosomes in body cells occur in <u>pairs</u>. For example, a human body cell with 46 chromosomes has 23 pairs of chromosomes. One member of each pair can be traced back to the father of the individual, having been contributed through the sperm cell. The other member of the chromosome pair can be traced back to the egg cell contributed

by the mother. Thus in the corn root tip cell shown in frame 26, the 20 chromosomes shown actually represent how many chromosome pairs? ____________________

- - - - - - - - - - - - - - - - - -

10

38. If we microscopically examine a root tip cell of a tomato plant, we would expect to count 24 chromosomes.

    a. How many pairs of chromosomes are in each cell of the tomato plant? ____________________

    b. How many chromosomes are in each gamete that gives rise to the mature tomato plant? ____________________

- - - - - - - - - - - - - - - - - -

a. 12; b. 12

39. The two members of a chromosome pair are similar in at least three respects: <u>size</u>, <u>shape</u>, and <u>type of gene content</u>. Such paired chromosomes are called <u>homologous</u> chromosomes. One member of each pair of homologous chromosomes contains genes contributed by the female parent, while the other member of the chromosome pair contains genes for the same traits contributed by the male parent. For example, if a particular human chromosome contains a gene affecting eye color, the homologous chromosome will also carry a gene affecting eye color.

    On the next page is a diagram of the paired chromosomes taken from a human cell. From this diagram it is easy to see that the two homologous chromosomes of each pair are similar in at least what <u>two</u> respects? ______________________________________

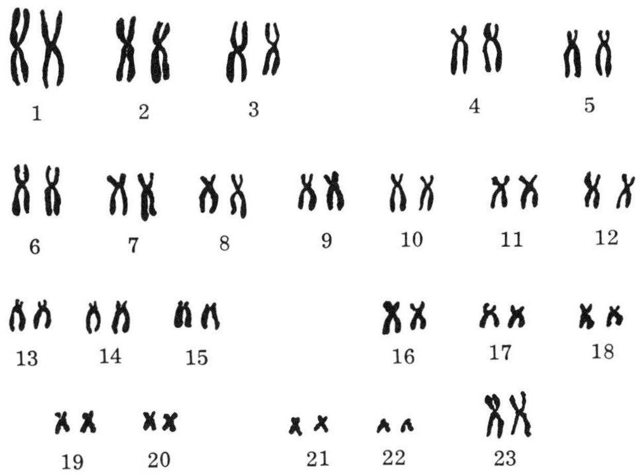

1 2 3 4 5

6 7 8 9 10 11 12

13 14 15 16 17 18

19 20 21 22 23

- - - - - - - - - - - - - - - - - - -

size and shape. From such a diagram it is impossible to learn anything about similarity in gene content.

40. How many different kinds of chromosomes occur in the diagram of chromosomes shown in frame 39? ____________________

- - - - - - - - - - - - - - - - - - -

23, since there are 23 different pairs of homologous chromosomes

41. In summary, the chromosomes that are members of a pair are called ____________________ chromosomes. One member of each pair can be traced back to the ____________________ of the individual and the other member of the pair can be traced back to the ________________ of the individual. Members of a chromosome pair are similar in at least what three respects? ____________________________________

____

____

- - - - - - - - - - - - - - - - - - -

homologous
father or male parent; mother or female parent (either order)
size, shape, type of gene content

42. Recall that meiosis is a nuclear division process that leads to the production of haploid nuclei. Thus, a human sperm has 23 chromosomes rather than the diploid number of 46. A human gamete—egg or sperm—does not have just <u>any</u> 23 chromosomes, however. Rather the process of meiosis occurs in such a precise and regular manner that each gamete receives one chromosome of each type—that is, one member of each homologous pair of chromosomes.

In the same way, when meiosis occurs in a corn plant, the resulting (haploid/diploid) ____________________ cells contain one (chromosome/gamete) ____________________ of each of the 10 different pairs.

- - - - - - - - - - - - - - - - - -

haploid; chromosome

43. In corn, then, the diploid chromosome number is restored in the process called ____________________, with each parent contributing one member of each of the 10 different pairs of (genes/chromosomes/gametes) ____________________.

- - - - - - - - - - - - - - - - - -

fertilization; chromosomes

44. Let's summarize these ideas with a brief exercise in which you match numbered items with definitions or descriptions.

_______ a. Process which restores the $2\underline{n}$ condition.

_______ b. Word used to describe the chromosomes that constitute a pair.

_______ c. Process which reduces chromosome number from $2\underline{n}$ to $\underline{n}$.

1. diploid
2. meiosis
3. analogous
4. homologous
5. chromosomes
6. genes
7. fertilization
8. DNA
9. haploid

_______ d. Darkly staining bodies found in a cell nucleus which bear hereditary material from cell to cell and from parent to offspring.

_______ e. Number of chromosomes characteristic of gametes.

10. gametes
11. correct answer not given

- - - - - - - - - - - - - - - - - - -

a. 7; b. 4; c. 2; d. 5; e. 9;

45.

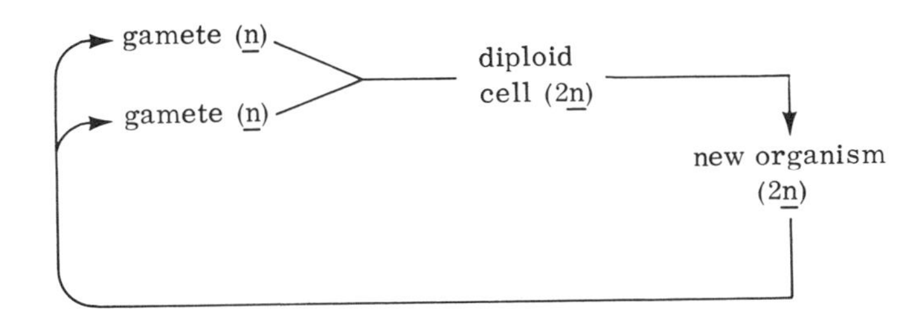

The word diagram above briefly summarizes sexual reproduction as it occurs in most animals. The animal—a dog, for example—is diploid (2$\underline{n}$ = 78 for the dog). The animal produces gametes, eggs in the female and sperms in the male, which are haploid, containing 39 chromosomes. These gametes or sex cells are produced by the process of meiosis. The diploid condition is reestablished when the gametes unite at the time of fertilization.

One detail has been omitted, however. How do we get from the single diploid cell—the fertilized egg or zygote that results from the union of gametes—to the multicellular organism, all the cells of which are diploid? The answer to that question is mitosis, a process of nuclear division that maintains a constant chromosome number in the daughter nuclei. Mitosis, unlike meiosis, does not separate pairs of homologous chromosomes at the time of cell division. The resulting daughter cells will therefore contain the exact numbers and kinds of chromosomes found in the parent cell. Mitosis results in increased cell number and is the basis for growth of the organism. If we can represent meiosis with the symbols $2\underline{n} \longrightarrow \underline{n}$, then we could represent mitosis with the symbols $2\underline{n} \longrightarrow$ _______.

- - - - - - - - - - - - - - - - - - -

2$\underline{n}$. In mitosis a diploid cell divides giving more diploid cells.

46. Thus, the animal life cycle shown in the word diagram in frame 45 might be represented more specifically as follows:

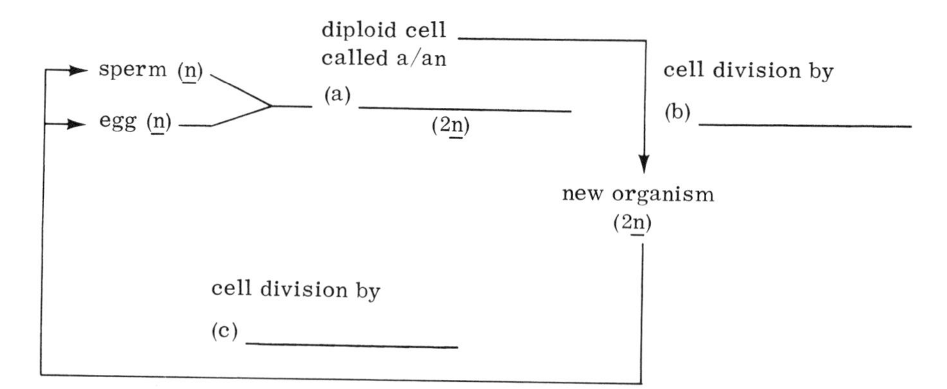

Complete the diagram by filling blanks a, b, and c with the proper words.

- - - - - - - - - - - - - - - - - -

a. zygote or fertilized egg; b. mitosis; c. meiosis

47. In humans meiosis leads to the formation of haploid cells called gametes, while fertilization restores the diploid condition in a cell called a zygote. If meiosis can be represented as $2\underline{n} \longrightarrow \underline{n}$, then we can use $2\underline{n} \longrightarrow 2\underline{n}$ to symbolize ________________.

- - - - - - - - - - - - - - - - - -

mitosis

48. Using the same approach, we could use $\underline{n} + \underline{n} = 2\underline{n}$ as a way to symbolize the process called ________________.

- - - - - - - - - - - - - - - - - -

fertilization

49. Following fertilization, how is the diploid chromosome number maintained so that all of the cells of the offspring have the diploid number of chromosomes? ____________________

- - - - - - - - - - - - - - - - - -

by the process of mitosis

50. How do these processes of meiosis, fertilization, and mitosis relate to plant reproduction cycles? Many simple plants, such as algae and fungi, have the potential of reproducing both sexually and asexually. The amount of food stored in the cells of the plant together with external environmental factors will determine whether the plant will reproduce sexually or asexually. In these simple plants the most common form of asexual reproduction involves the production by mitosis of one- or two-celled structures called spores. Such spores are capable of giving rise to new individuals identical to the parent plant. Thus, spores achieve the same end result seen in the forms of asexual reproduction mentioned for the African violet or the starfish—the offspring is (similar to/different from/identical to) ____________ ____________ the parent.

- - - - - - - - - - - - - - - - - -

identical to

51. Let's look at an example of this form of asexual reproduction. A piece of bread is covered with cottony black mold. You examine a bit of the sooty, powdery material under the microscope and find countless numbers of tiny cells. Further investigation reveals that each cell has been produced by mitosis and that it readily gives rise to a new mold plant. You should probably conclude that these cells are (zygotes/spores) ____________ and that they represent a means of (sexual/asexual) ____________ reproduction.

- - - - - - - - - - - - - - - - - -

spores; asexual. If this black mold is grown under somewhat different conditions of nutrition and environment, you may discover that it also reproduces sexually.

52. Many times the individual plants of a given species reproduce in only one fashion. Thus, a particular plant may reproduce solely by gamete formation, while another plant may reproduce only by special spores which are produced by meiosis.

Ulva, or sea lettuce, may be picked up along the seashore where it has washed in with the tide. Slightly resembling a leaf of lettuce, any one "leaf" is capable of reproducing by only one method. Thus, some Ulva "leaves" produce sex cells or ____________________, while others produce special ____________________ by the process of meiosis.

\- - - - - - - - - - - - - - - - - -

gametes; spores

53. Ulva plants that produce gametes may logically be called gamete plants and those which produce spores may be called spore plants. Indeed, they are so designated, but by terms derived by combining Greek words to convey these meanings. Gamete-forming plants are thus termed gametophytes [gameto + phyta (plants)], while spore-producing plants are called sporophytes [sporo + phyta (plants)]. A given plant species includes both gametophyte and sporophyte phases in its total life cycle.

All sea lettuce (Ulva) plants are identical to each other in outward appearance to the naked eye. If, however, a microscopic examination were made of the reproductive cells along the edges of many sea lettuce plants, one would observe that some plants produce only spores, while others produce only gametes.

What term can be applied to the spore-producing plant?

____________________ To the gamete-producing plant?

____

\- - - - - - - - - - - - - - - - - -

sporophyte; gametophyte

54. At this point you may be wondering why both sporophytes and gametophytes occur in a plant population and what the relationship between them may be. When the gametes produced by gametophytes fuse, the resulting cell, the zygote, develops into a plant that is a sporophyte! The zygote, then, is the first cell of the sporophyte generation.

The mature sporophyte produces spores which develop into gametophytes! A spore, then, constitutes the first cell in the gametophyte generation. In other words, a parent produces an offspring unlike itself in its method of reproduction. Sporophytes give rise to gametophytes which in turn give rise to more sporophytes.

Check the statement(s) that are correct.

_______ a. Gametophytes form offspring that reproduce by spores.

_______ b. Germinating spores form sporophytes.

_______ c. Sporophytes give rise to spores that develop into gametophytes.

_______ d. When gametes unite they form a gametophyte.

_______ e. The zygote is the first cell in the sporophyte generation.

- - - - - - - - - - - - - - - - - -

a, c, and e

55. The predictable pattern of gametophyte-sporophyte-gametophyte is called an alternation of generations and is typical of almost all plant life cycles.

In the following word diagram of alternation of generations, insert the letters of the terms into the appropriate spaces within the parentheses.

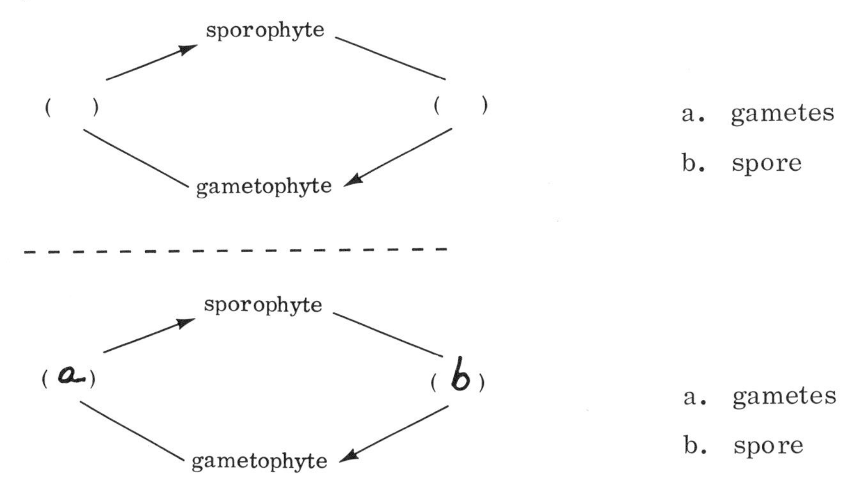

56. Earlier you learned that fertilization or union of gametes results in a double (diploid) set of chromosomes in the zygote. You also learned that meiosis in the life cycle of sexually reproducing organisms results in a single (haploid) chromosome set before fertilization again occurs.

Since alternation of generations involves sexual reproduction, both fertilization and meiosis occur within the life cycle. In the alternation of generations life cycle, fertilization occurs when the gametes unite. Meiosis, on the other hand, occurs in the structure which produces the spores. In other words, spores are formed by meiosis.

In the following word diagram of an alternation of generations, write "meiosis" and "fertilization" in the appropriate places.

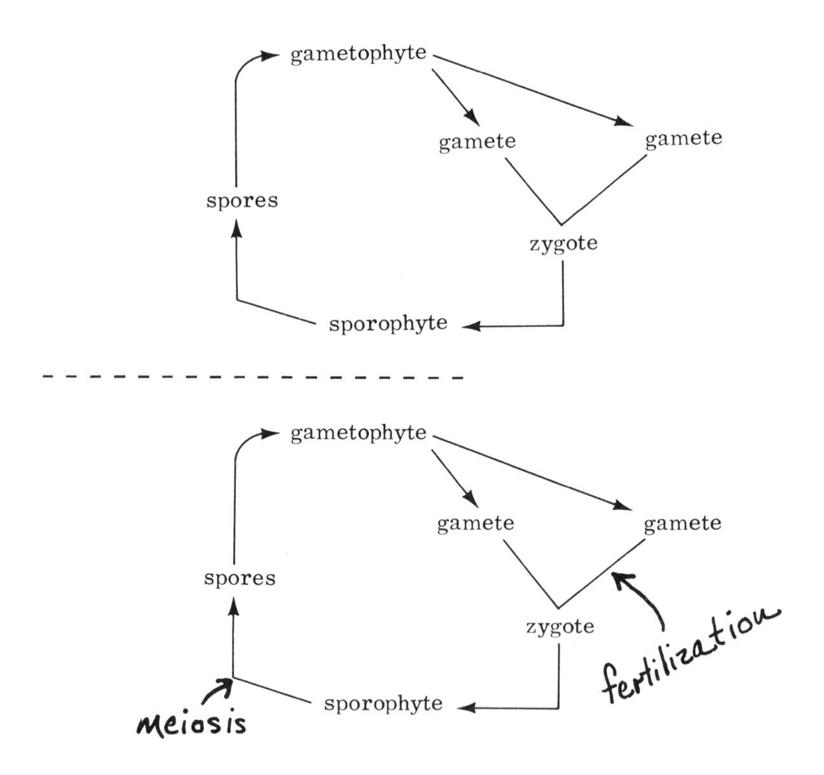

57. Thus, in alternation of generations a 2n plant alternates with an n plant. In the following word diagram of alternation of generations, write "2n" and "n" in the appropriate spaces marked by parentheses.

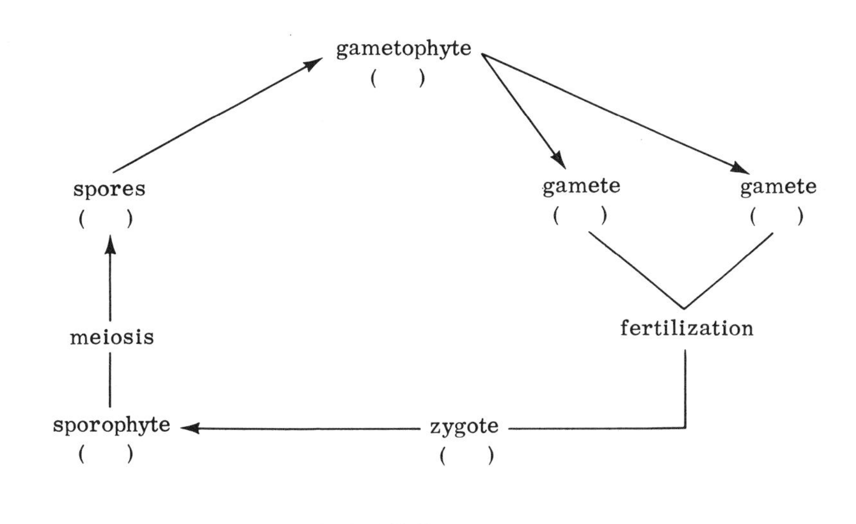

- - - - - - - - - - - - - - - - - -

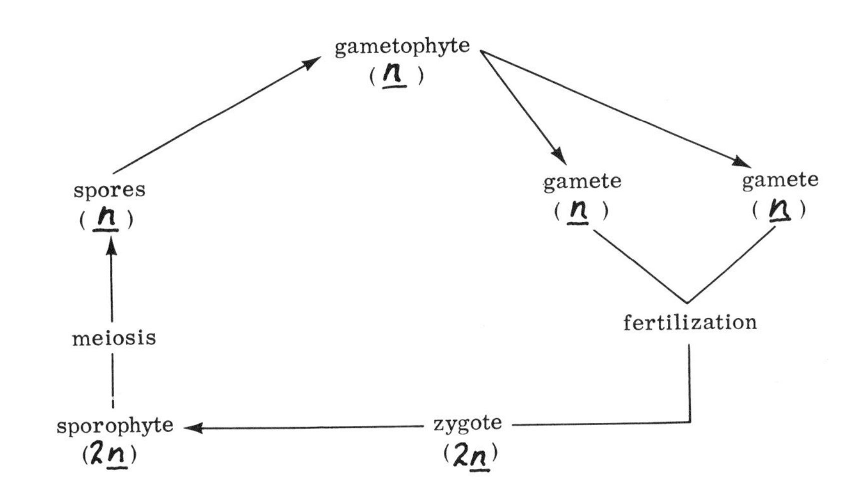

58. The diploid chromosome number in the common garden pea is 14. How many chromosomes will be found in each of the following kinds of cells in the pea plant?

_______ a. spore

_______ b. sporophyte

_______ c. gamete

_______ d. zygote

_______ e. gametophyte

- - - - - - - - - - - - - - - - - -

a. 7; b. 14; c. 7; d. 14; e. 7

59. To summarize the main features of alternation of generations, match the following words and phrases.

_______ a. generation which forms cells by meiosis

_______ b. 2n generation

_______ c. gamete-producing generation

_______ d. the zygote is the first cell of this generation (from the standpoint of chromosome number) number)

_______ e. the generation terminated by fertilization

1. gametophyte generation
2. sporophyte generation

- - - - - - - - - - - - - - - - - -

a. 2; b. 2; c. 1; d. 2; e. 1

60. Similarly, match these words and phrases.

_______ a. spores

_______ b. mature sporophyte

_______ c. zygote

_______ d. gametes

_______ e. mature gametophyte

1. diploid generation
2. haploid generation

- - - - - - - - - - - - - - - - - -

a. 2; b. 1; c. 1; d. 2; e. 2

61. Complete the following diagram of a plant life cycle possessing an alternation of generations by inserting the missing words.

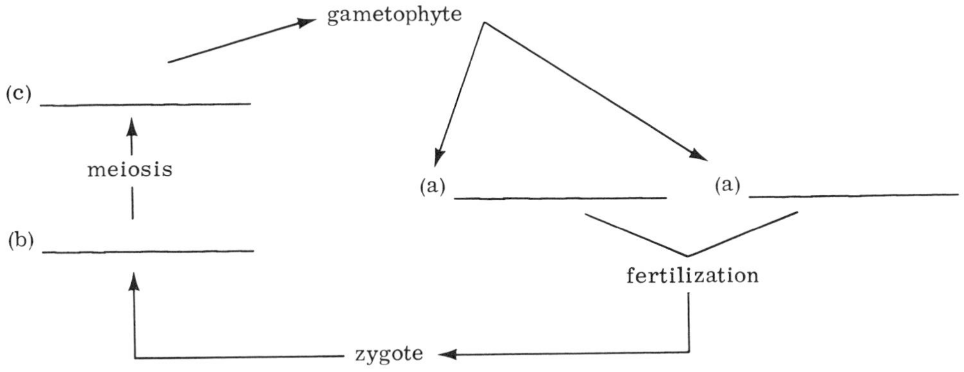

gametophyte

(c) ____________

meiosis

(a) ____________ (a) ____________

(b) ____________

fertilization

zygote

- - - - - - - - - - - - - - - - - - -

a. gametes; b. sporophyte; c. spores

62. Listed below are some stages and processes that occur in alternation of generations. Put these lettered items in correct chronological order to complete the life cycle.

    (A) zygote, (B) gametophyte, (C) meiosis, (D) gametes, (E) spores, (F) sporophyte, (G) fertilization.

- - - - - - - - - - - - - - - - - - -

Since the process is cyclic and you can begin at any point, any one of the following sequences is correct:

A-F-C-E-B-D-G-A
B-D-G-A-F-C-E-B
C-E-B-D-G-A-F-C
D-G-A-F-C-E-B-D
E-B-D-G-A-F-C-E
F-C-E-B-D-G-A-F
G-A-F-C-E-B-D-G

SELF-TEST
Sexual and Asexual Reproduction

Before you examine how alternation of generations fits into the life cycles of specific kinds of plants, check your progress in learning by taking this short self-test. All questions should be answered in the light of information presented in Chapter 1 of this book. Answers are given at the end of this quiz.

1. In animals, the process of meiosis results in the formation of cells called ____________________, while in plants meiosis usually results in the production of cells called ____________________.

2. Uniformity of type is characteristic of (sexual/asexual) ____________ reproduction, while variety among the offspring results from ____________________ reproduction.

3. Listed below are a number of stages and processes that occur in alternation of generations. Put these lettered items in correct chronological order by letter. Begin at any place you desire, proceeding in correct order to complete the life cycle.

   (A) zygote, (B) gametophyte, (C) meiosis, (D) gametes, (E) spores, (F) sporophyte, (G) fertilization.

   ____

4. Identify each of the processes symbolized below.

   $\underline{n} + \underline{n} = 2\underline{n}$ ____________________

   $2\underline{n} \longrightarrow 2\underline{n}$ ____________________

   $2\underline{n} \longrightarrow \underline{n}$ ____________________

5. In root tip cells which are part of the sporophyte generation of a cabbage plant, one can count 18 chromosomes.

   a. If one of these root tip cells divides to form two cells, how many chromosomes will be found in the daughter root tip cells? ______

b. How many pairs of homologous chromosomes occur in these cabbage root tip cells? _______

c. How many chromosomes would be present in an egg cell in a cabbage plant? _______

d. A sperm cell? _______

e. A zygote? _______

6. Homologous chromosomes are similar in at least what three ways?

____

7. In the mouse the $2\underline{n}$ chromosome number is 40. Using this information, answer the following questions.

a. How many different kinds of chromosomes occur in mouse cells?

____

b. What is the total number of chromosomes in a mouse zygote?

____

c. What is the total number of chromosomes in a mouse sperm?

____

d. How many different kinds of chromosomes are there in a mouse egg? _______

8. In alternation of generations, name the following.

a. The <u>first</u> cell in the sporophyte generation. _______________

b. All of the <u>haploid</u> stages. ______________________________

c. Cells produced by meiosis. _________________

d. The type of nuclear division by which the zygote develops into the sporophyte. _________________

9. Suppose you found a certain simple plant species which reproduced only by asexual means. Which of the following might be expected to be true of this plant?

_______ a. All offspring of a single plant specimen would be genetically identical.

_______ b. This plant species would probably be adapted to a wide variety of environments.

_______ c. Meiosis would give rise to spores in this plant species.

_______ d. Alternation of generations could be demonstrated in this plant species.

10. Match items with their descriptions.

_______ a. characterized by physiologically different gametes

_______ b. characterized by morphologically identical gametes

1. heterogamy (Oedogonium)
2. isogamy (Ulothrix)

Answers

1. gametes; spores (frames 1, 30, 56)

2. asexual; sexual (frames 6, 7, 11, 12)

3. Since the process is cyclic and you can begin at any point, any one of the following sequences is correct.

   A-F-C-E-B-D-G-A
   B-D-G-A-F-C-E-B
   C-E-B-D-G-A-F-C
   D-G-A-F-C-E-B-D
   E-B-D-G-A-F-C-E
   F-C-E-B-D-G-A-F
   G-A-F-C-E-B-D-G

   (frames 54-62)

4. fertilization; mitosis; meiosis (frames 45, 47, 48)

5. a. 18 (frame 45)
   b. 9 (frames 37-41)
   c. 9 (frame 58)
   d. 9 (frame 58)
   e. 18 (frame 58)

6. size, shape, type of gene content (any order) (frame 39)

7. a. 20 (frame 40)
   b. 40 (frame 35)
   c. 20 (frame 35)
   d. 20 (frame 40)

8. a. zygote (frame 59)
   b. spores, gametophyte, gametes (frame 60)
   c. spores (frame 56)
   d. mitosis (frame 46)

9. a. true (frame 6)
   b. false (frames 8-10)
   c. false (frame 34)
   d. false (frame 56)

10. a. 1 and 2 (frame 20)
    b. 2 (frame 20)

CHAPTER TWO

Alternation of Generations—Patterns of Plant Reproduction

In this chapter you will learn how the generalized description of alternation of generations you studied in Chapter 1 applies to the life cycles of specific plants. We will begin by looking at the life cycle of a moss plant because alternation of generations is easy to understand in moss.

1. In a few plants such as the sea lettuce *Ulva*, the sporophyte and gametophyte generations are indistinguishable from each other in form. However, in most plants with an alternation of generations in the life cycle, sporophytes and gametophytes differ widely in appearance. Whether alike or different outwardly, gametophytes and sporophytes differ from each other in chromosome number and in manner of reproduction.

   Thus we may conclude that form or appearance is the (most/least) ____________________ important criterion for distinguishing between sporophyte and gametophyte generations.

- - - - - - - - - - - - - - - - - - - -

least

2. Suppose you were given two plants that appeared superficially to be identical. You were asked to determine whether one might, in fact, be a gametophyte and the other a sporophyte. Suggest two kinds of evidence that might enable you to ascertain which was gametophyte and which was sporophyte.

   (1) ____

   (2) ____

- - - - - - - - - - - - - - - - - - - -

(1) chromosome number—sporophyte is 2n; gametophyte is n
(2) manner of reproduction—sporophyte produces spores by meiosis; gametophyte reproduces by union of gametes

3. A moss is an excellent example of a plant whose life cycle includes an alternation of generations with dissimilar sporophyte and gametophyte forms. If you were to remove a clump of moss from a decaying log and examine it carefully, you would find that the clump is composed of many individual tiny green, leafy plants. Examining the tips of these plants microscopically, you would find eggs and sperms present in special gamete-producing structures.

We could conclude, then, that the "old moss-covered bucket that hung in the well" was overgrown with (sporophyte/gametophyte) ____________________ plants whose chromosome numbers were (diploid/haploid) ____________________.

- - - - - - - - - - - - - - - - - - -

gametophyte; haploid

4. Sexual reproduction in mosses occurs only when water is present. Moss sperms possess special organelles called flagella which allow them to swim to the eggs lying in special structures at the tips of the leafy gametophytes. There fertilization occurs. Because water must be present to provide a medium through which the sperm can reach the ____________________, a moss plant growing in a dry place cannot reproduce (sexually/asexually) ____________________.

- - - - - - - - - - - - - - - - - - -

egg; sexually

5. Moss gametophytes are delicate plants composed of thin layers of cells. Unlike most plants which grow out of water, most moss gametophytes lack a protective waxy covering, so they dry out quickly in sunny, warm, dry air. Moss gametophytes commonly grow close to the ground in wet, shady habitats. This allows the sperms to swim to the egg and prevents the death of the gametophytes as a result of drying out. The eggs of moss remain in place at the tips of the leafy gametophytes where fertilization and zygote formation occur.

What two problems of survival would need to be met by a moss plant if it were to grow in a sunny dry area? ____________________

____

____

- - - - - - - - - - - - - - - - - -

The moss plant would need a watery medium in which the sperm could swim to the egg and fertilize it. Secondly, it would need sufficient protection from drying out.

6. The diploid zygote divides by the process of mitosis and immediately develops into a diploid sporophyte which is <u>attached to</u> the parent gametophyte.

   From the information just given, determine which of the following statements are true. Place a check mark before the true statements.

   ________ a. The sporophyte is attached to the gametophyte.

   ________ b. A diploid plant is attached to a haploid parent plant.

   ________ c. The sporophyte was formed from the zygote.

- - - - - - - - - - - - - - - - - -

All three statements are true.

7. All food and water required by the sporophyte are derived from the gametophyte. The moss sporophyte remains <u>permanently attached</u> to the parent gametophyte, not only in its initial development, but also for the remainder of its life.

   Can a moss sporophyte live without the moss gametophyte?

   __________ Why? ____

   ____

- - - - - - - - - - - - - - - - - -

No, the moss sporophyte is attached to and dependent on the moss gametophyte.

8. As the dependent moss sporophyte develops, it assumes an appearance quite different from the green, leafy parent gametophyte. The sporophyte forms a long slender stalk (the seta) at the tip of which is borne a swollen bulblike portion (the capsule).

   Label the parts of the moss sporophyte as depicted in the drawing below.

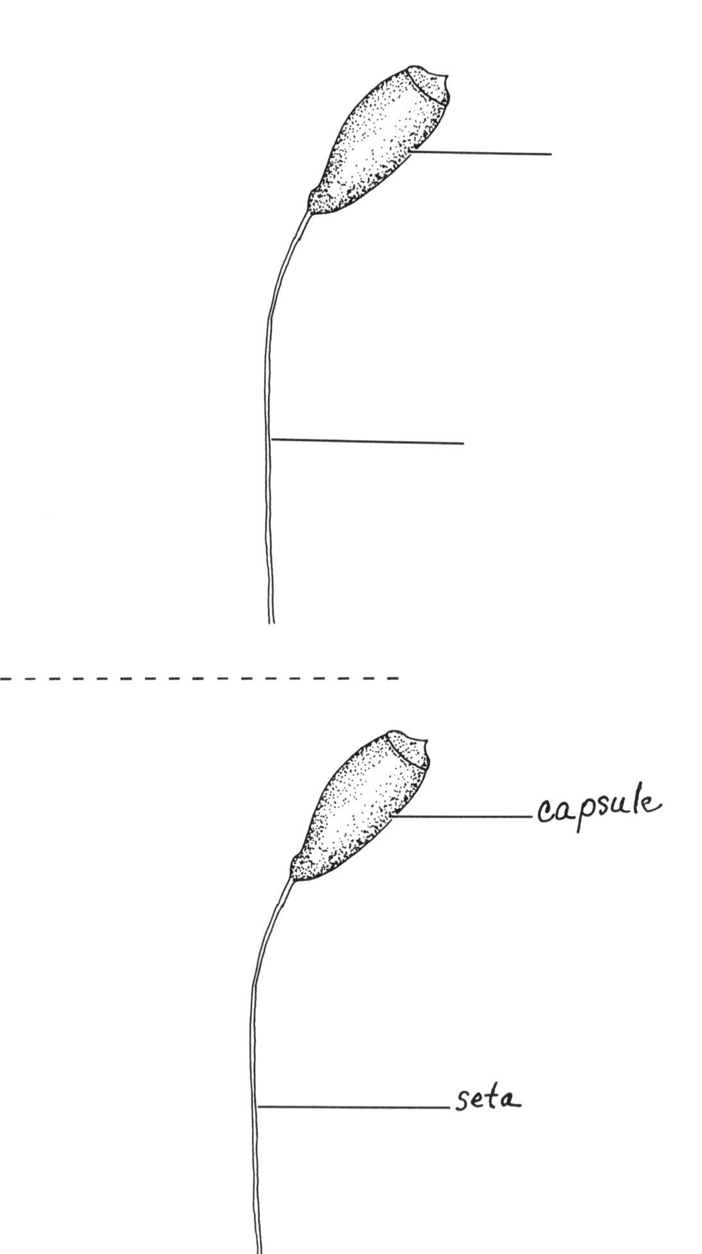

9. The moss sporophyte is attached to the gametophyte generation, as shown in the drawing below. From what you have learned so far, label the gametophyte and sporophyte generations.

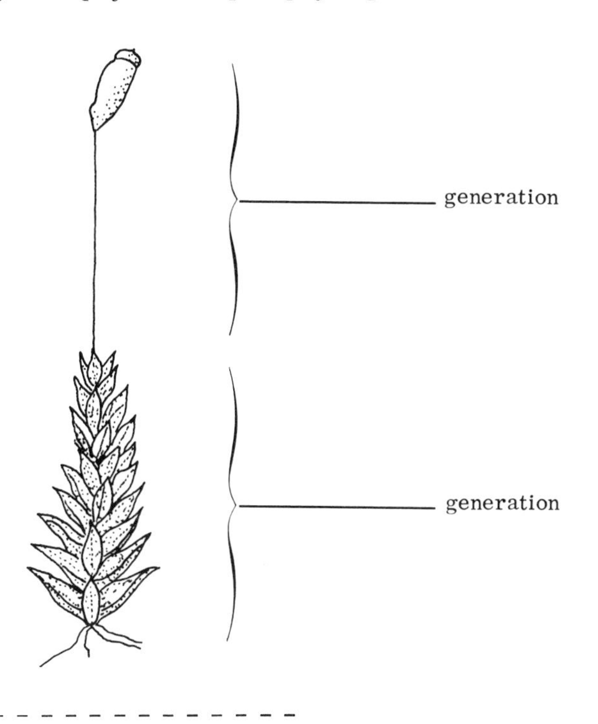

- - - - - - - - - - - - - - - - - - - -

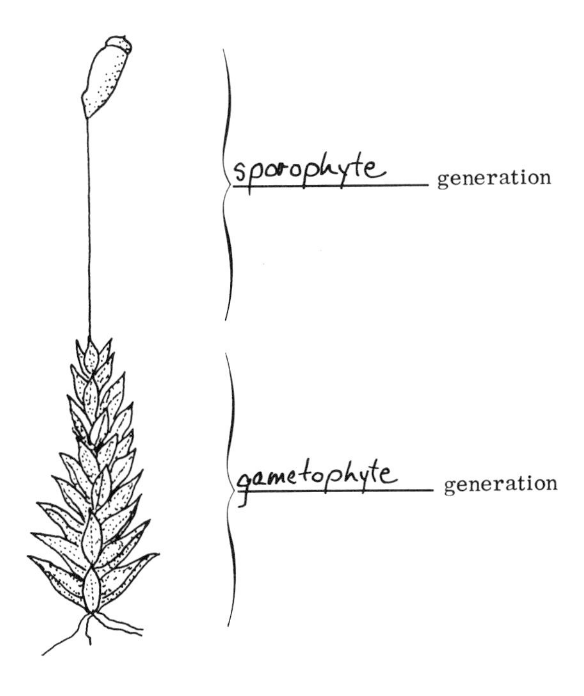

10. The chief function of the moss sporophyte capsule is to produce spores by the process of meiosis. As the capsule matures, a covering falls away to allow the release of hundreds of spores into the air.

    Since the spores are produced by the process of (mitosis/meiosis/fertilization) ____________________, they can be expected to contain the (n/2n/3n) _______ number of chromosomes.

- - - - - - - - - - - - - - - - - - -

meiosis; n

11. As you learned in Chapter 1, frame 54, which dealt with the nature of alternation of generations, gametophytes produce gametes that fuse and give rise to sporophytes and sporophytes produce spores that germinate (sprout) to form gametophytes.

    Germination of the moss spores formed by the sporophyte generation produces plants of the (gametophyte/sporophyte) ______________ generation which bear ("leaves"/capsules) ____________________.

- - - - - - - - - - - - - - - - - - -

gametophyte; "leaves"

12. In Chapter 1, frame 56 (and again in frame 10 of this chapter), you learned that the spores produced by the sporophyte are formed by meiosis when alternation of generations occurs.

    In the life cycle of a moss, meiosis occurs in the (capsule/seta/leafy gametophyte tips) ____________________ and forms (spores/gametes) ____________________.

- - - - - - - - - - - - - - - - - - -

capsule; spores

13. In summary, the moss life cycle involves a leafy haploid gametophyte reproducing by gametes to produce a non-leafy, dependent sporophyte whose spores give rise to gametophytes.

    In the diagram on the next page, label the gametophyte and sporophyte generations and insert the words "fertilization" and "meiosis" in the appropriate places to indicate where these processes occur.

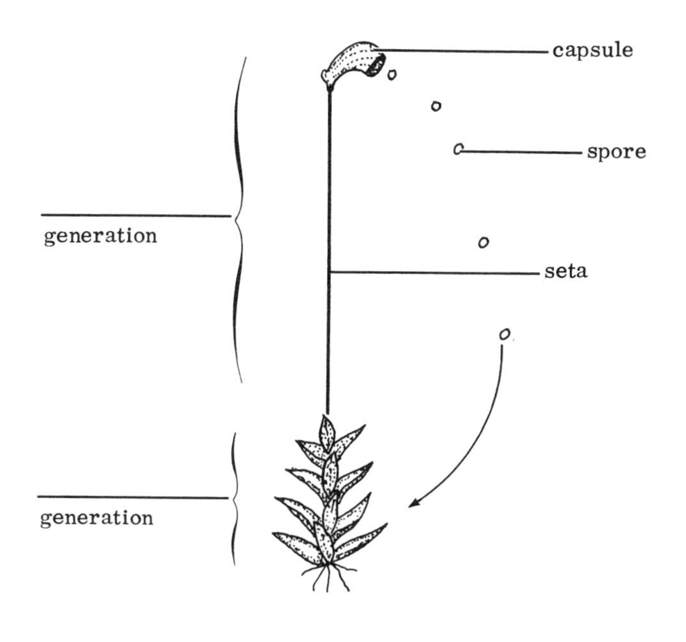

- - - - - - - - - - - - - - - - - -

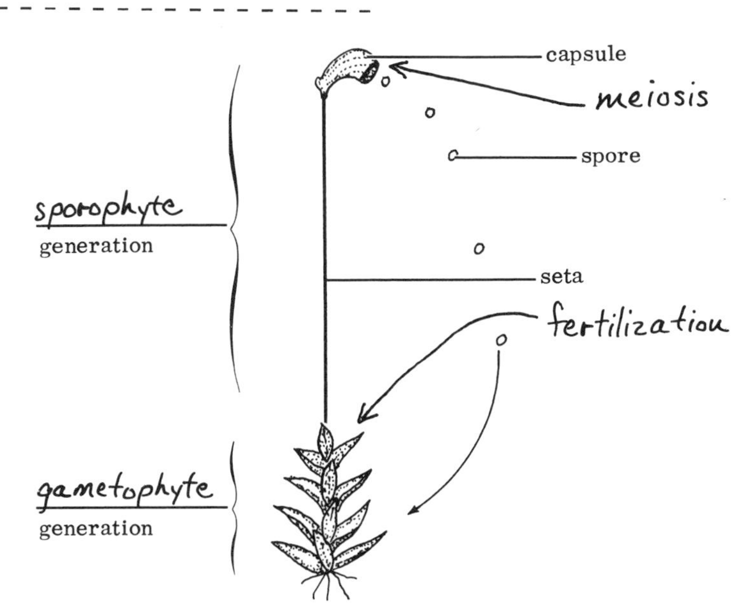

14. In frame 13, you indicated where meiosis and fertilization occur in the moss life cycle. On the basis of that cycle, which of the following are haploid (n) and which are diploid (2n)?

_______ a. seta

_______ b. spores

_______ c. gametophyte generation

_______ d. capsule

_______ e. sporophyte generation

- - - - - - - - - - - - - - - - - -

a. 2n; b. n; c. n; d. 2n; e. 2n

15. Now, in a short paragraph, summarize the life cycle of a moss, describing how alternation of generations occurs by using words such as haploid, diploid, sporophyte, gametophyte, capsule, seta, meiosis, mitosis, fertilization, spores, gametes, etc.

- - - - - - - - - - - - - - - - - -

Sample answer:

The green leafy moss plant is the gametophyte generation. As such, it reproduces by forming gametes—sperms and eggs. The haploid sperm swims in a drop of water to the haploid egg at the tip of the leafy gametophyte. Fertilization takes place when the sperm and egg unite to form a diploid fertilized egg or zygote, which divides by the process of mitosis to form the diploid sporophyte generation. The sporophyte consists of a slender stalk, the seta, and a larger capsule at the tip of the seta. The process of meiosis takes place in the capsule and forms haploid spores which drop to the ground, germinate, and give rise to new leafy moss plants of the haploid gametophyte generation.

16. As another example of a plant with an alternation of generations, we will study Marchantia, a kind of liverwort. The liverworts are considered to be closely related to the mosses because their life cycles are similar. Your understanding of the moss life cycle should be reinforced by studying Marchantia.

    Marchantia forms a narrow, green, ribbonlike plant body on damp rocks and soil. Since this stage in the life cycle is comparable to the green, leafy generation of the moss, what generation do you think this is? ____________________ What method of reproduction would you expect it to have? ____________________

\- - - - - - - - - - - - - - - - - - -

gametophyte; sexual, by gamete formation

17. When the green plant of Marchantia reproduces, it forms special branches on its upper surface (see diagrams).

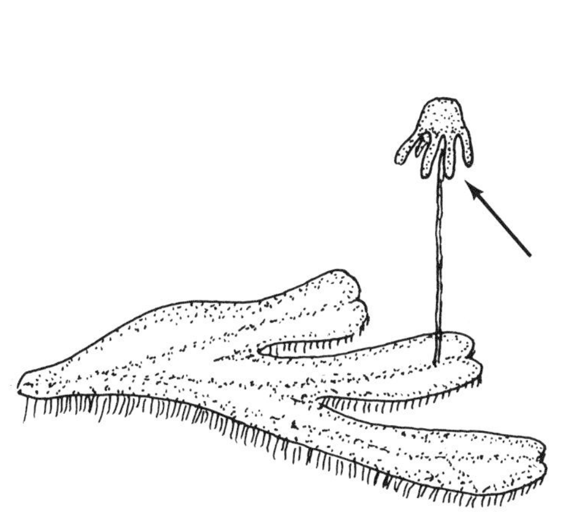

A. Marchantia with a "female" branch

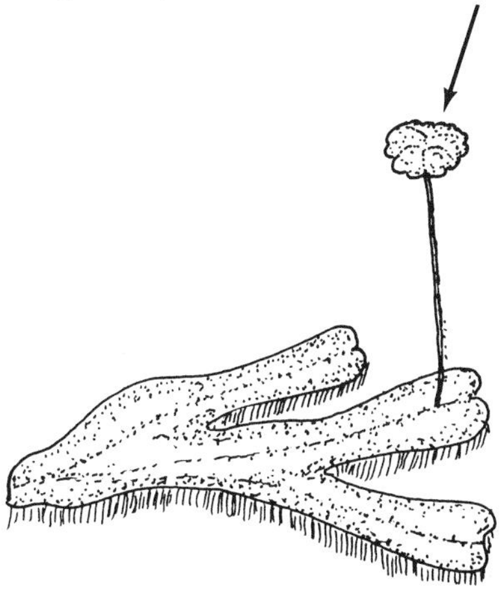

B. Marchantia with a "male" branch

We can expect to find the female gametes, or eggs, on the underside of the umbrellalike branch marked with an arrow in figure A, and male gametes, or sperms, on the upperside of the disklike branch marked with an arrow in figure B.

When fertilization occurs, on which branch will the zygote be formed? ____________________

- - - - - - - - - - - - - - - - - - -

"female" branch

18. Since the zygote represents the first stage of the (gametophyte/sporophyte) ____________________, one would look on the lower side of the "female" branch to find the mature (sporophyte/gametophyte) ____________________ generation. Like its relative the moss, the sporophyte of the liverwort is dependent on the gametophyte. Therefore, the dependent generation of <u>Marchantia</u> will obtain its food from the ("male"/"female") ____________________ branch of the (sporophyte/gametophyte) ____________________ generation.

- - - - - - - - - - - - - - - - - - -

sporophyte; sporophyte; "female"; gametophyte

19. The sporophytes of moss and <u>Marchantia</u> differ from each other chiefly in size and complexity. As in the moss, the spore-producing structure of <u>Marchantia</u> is the capsule. Meiosis, therefore, occurs in the capsule of <u>Marchantia</u> and leads to the production of haploid cells called spores.

    Now let's summarize the <u>Marchantia</u> alternation of generations life cycle by placing the following stages and processes in correct chronological order by letter. Begin at any place you desire, proceeding in correct order to complete the life cycle. Remember that the <u>Marchantia</u> life cycle is very much like that of a moss.

    (A) <u>Marchantia</u> gametophyte, (B) <u>Marchantia</u> sporophyte, (C) gametes, (D) spores, (E) fertilization, (F) meiosis, (G) zygote.

- - - - - - - - - - - - - - - - - - -

C → E
A G
D ← F ← B

20. The sporophyte of Marchantia, unlike that of the moss, remains quite small and almost hidden by gametophytic tissue (see diagram).

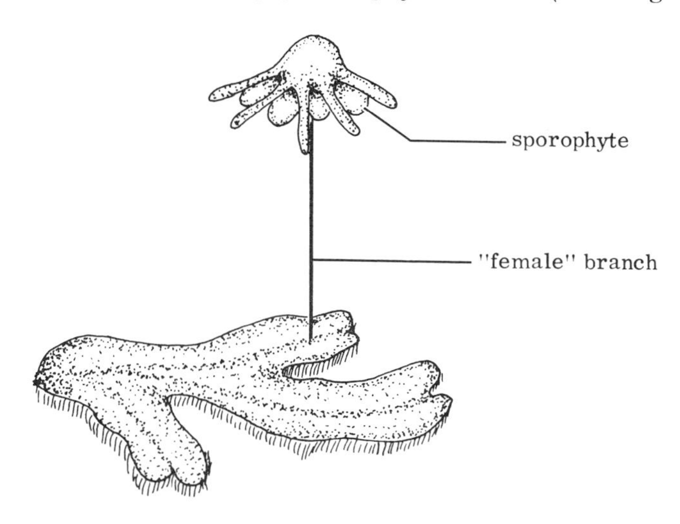

Marchantia gametophyte with sporophytes

The smaller generation of Marchantia is the (gametophyte/sporophyte) ____________________ which is virtually hidden by the ("female"/"male") ____________________ branch.

- - - - - - - - - - - - - - - - - - - -

sporophyte; "female"

21. To review alternation of generations as illustrated by the mosses and the liverworts, match the items on the following page.

1. liverwort sporophyte
2. moss sporophyte
3. liverwort gametophyte
4. moss gametophyte

_______ a. } conspicuous generation
_______ b.

_______ c. } meiosis brings them into existence
_______ d.

_______ e. smaller liverwort generation

_______ f. leafy generation

_______ g. } form reproductive cells by mitosis
_______ h.

_______ i. } arise as the result of spore germination
_______ j.

_______ k. } capsules present
_______ l.

_______ m. } obtain food from parent
_______ n.

- - - - - - - - - - - - - - - - - -

a. 3, b. 4 (either order); c. 3, d. 4 (either order); e. 1; f. 4; g. 3, h. 4 (either order); i. 3, j. 4 (either order); k. 1, l. 2 (either order); m. 1, n. 2 (either order)

22. Recall from frame 4 of this chapter that water must be present so that fertilization can occur in mosses. The same is also true for Marchantia. Thus, a form of plant body which grows close to the surface on wet soil is (advantageous/disadvantageous) _______________ to the gametophyte generation.

- - - - - - - - - - - - - - - - - -

advantageous

23. The low growth of a plant body which reproduces by gametes is not advantageous to a plant which forms spores, since spores are more efficient agents of reproduction when they are scattered over a wide area. Thus, a form of plant body which grows to a considerable

height above the ground might be expected to be (advantageous/disadvantageous) ____________________ to the sporophyte generation.

- - - - - - - - - - - - - - - - - - -

advantageous

24. With the exception of algae, fungi, mosses, and liverworts, more complex plant groups have well developed sporophytes which are adapted to a true land-dwelling or terrestrial existence. The height of the sporophyte permits more efficient spore dispersal. Other important adaptations, such as the presence of a waxy covering and special tissues for upward and downward movement of water and food, are also necessary for terrestrial existence. In fact, such adaptations are considered to be major advances in the evolutionary development of the plant kingdom. Many biologists conclude, therefore, that the most primitive plants were (aquatic/terrestrial)

____________________.

- - - - - - - - - - - - - - - - - - -

aquatic

25. A terrestrial plant species, if it is to reproduce, must be adapted to a habitat which allows for the union of gametes, while at the same time it must adapt a form of growth which favors efficient spore distribution.

As an advantage for living in a terrestrial habitat, the incorporation of an alternation of generations with a large (gametophyte/sporophyte) ____________________ generation would do much to insure the wide dissemination of (gametes/spores) ____________________ and increase the likelihood of survival for the plant.

- - - - - - - - - - - - - - - - - - -

sporophyte; spores

26. All green plants which are more complex than algae, fungi, liverworts, and mosses possess: (1) an alternation of generations, (2) a conspicuous sporophyte generation, and (3) an inconspicuous gametophyte generation.

The life cycle of a fern includes a large form and a tiny green form approximately one-quarter inch in diameter. The tiny plant represents the (sporophyte/gametophyte) ____________________ generation.

- - - - - - - - - - - - - - - - - - -

gametophyte

27. One of many adaptations shared by nearly all terrestrial sporophytes is the possession of special tissue for the transportation of water and food up and down the length of the plant. Such vascular tissue allows the sporophyte to attain spectacular heights, and makes possible the possession of true roots, stems, and leaves.

    The possession of true roots, stems, and leaves is a characteristic of the (sporophyte/gametophyte) ____________________. These organs have vascular tissue, the purpose of which is to ____________ ____.

- - - - - - - - - - - - - - - - - - -

sporophyte; conduct food and water within the plant

28. From the previous discussion of the nature of terrestrial plants, we can infer that the large leaves of a Boston fern belong to the ____________________ generation, and that they possess internal conducting cells called ____________________ tissue.

- - - - - - - - - - - - - - - - - - -

sporophyte; vascular

29. To summarize, terrestrial plants such as the Boston fern have a life cycle in which the sporophyte generation is large and conspicuous.

    a. What is the advantage of the large conspicuous sporophyte?

    ____

    ____

b. What other adaptation to terrestrial existence do these plants possess? ____________________

____

- - - - - - - - - - - - - - - - - - -

a. The large sporophyte, which is well above the surface of the soil, permits wide dissemination of spores and thus the survival of the species.
b. Vascular tissue that conducts food and water throughout the large sporophyte.

30. The spores of terrestrial sporophytes like those of some aquatic forms are produced in special spore containers called sporangia (singular sporangium) [spore + angium (covering or container)] .

Sporangia occur in all terrestrial sporophytes, though they are not always formed on the same organ of the plant. In a few instances, sporangia occur on the stems, but in most sporophytes the sporangia are produced on special leaflike structures.

In terrestrial plants one would expect to find spore containers called ____________ to be produced either on ____________ or on ____________.

- - - - - - - - - - - - - - - - - - -

sporangia; stems, leaflike structures (either order)

31. The sporangia-producing structures are termed sporophylls (literally spore-leaves). These structures resemble leaves and are probably derived from them.

Leaflike structures bearing sporangia are called ____________ and they occur on all vascular sporophytes which do not form sporangia on (leaves/stems/gametophytes) ____________.

- - - - - - - - - - - - - - - - - - -

sporophylls; stems

32. Sometimes sporophylls are identical in appearance to true leaves, but

most sporophylls are highly specialized and only slightly resemble leaves.

Brown dotlike structures often form on the leaves of the Boston fern. Which of the following statements are obviously erroneous explanation(s) for the fern's appearance?

_______ a. The plant probably is diseased.

_______ b. The leaves bearing the dotlike structures are actually sporophylls.

_______ c. The brown dots represent gamete-producing structures.

_______ d. The plant is reproducing by spores.

_______ e. Meiosis is occurring in the cells of the brown dots because these brown dots are clusters of sporangia.

_______ f. Sporophylls and leaves are essentially identical in appearance in the Boston fern.

_______ g. Sporangia are borne on leaves instead of sporophylls in the Boston fern.

- - - - - - - - - - - - - - - - - -

c and g are incorrect answers. Although a may seem reasonable to a non-biologist, the other choices (b, d, e, and f) are the correct explanations for these brown dotlike structures.

33. When sporophylls are present on the sporophyte they may be scattered at random among the leaves. Commonly, however, sporophylls are grouped together into more or less tight aggregations to form <u>cones</u> or <u>strobili</u> (singular, <u>strobilus</u>).

A strobilus or cone is composed of ______________________. Is it possible for a plant to possess sporophylls and lack strobili or cones?

____

- - - - - - - - - - - - - - - - - -

sporophylls; yes—the Boston fern is a good example

34. In Boston fern, the sporophylls are scattered among the leaves which they closely resemble in size and shape.

Which of the statements below accurately describes the condition in the Boston fern? Place a check mark before the true statements.

_______ a. Because the structures which bear sporangia are identical to the leaves in size and shape, they should not be termed sporophylls.

_______ b. Because many sporangia are clustered on a single sporophyll, the fern possesses cones.

_______ c. The fern lacks strobili.

_______ d. The fern lacks cones.

_______ e. Sporophylls are part of the sporophyte plant.

- - - - - - - - - - - - - - - - - -

c, d, and e

35. The club moss is a common small terrestrial plant found in many wooded areas. Although not a true moss but a vascular plant, the club moss is appropriately termed "club," for swollen club-shaped bodies are evident at the tips of many of its stems. Dissection of the "clubs" reveals that they are composed of a central stem to which are attached many small leaflike structures which bear sporangia on their upper surfaces.

From these statements, one may conclude that the leaflike sporangia-bearing structures are (leaves/sporophylls/strobili)

_____________________, that the cluster of the leaflike structures which form the "club" is a (capsule/sporophyll/strobilus)

_____________________, and that the plant must be the (gametophyte/sporophyte) _____________________ generation having the (haploid/diploid) _____________________ chromosome number.

- - - - - - - - - - - - - - - - - -

sporophylls; strobilus; sporophyte; diploid

36. What is the structural relationship between sporophylls, strobili, and sporangia? ____

____

- - - - - - - - - - - - - - - - - -

Sporangia are located on sporophylls; sporophylls are grouped to form strobili.

37. When the spores of Boston fern come in contact with the damp soil, they produce tiny, green heart-shaped structures which seldom exceed one quarter of an inch in diameter. Obviously, this generation resulting from the germination of a spore is the (sporophyte/gametophyte) ____________________ generation and possesses the (haploid/diploid) ____________________ chromosome number.

- - - - - - - - - - - - - - - - - - -

gametophyte; haploid

38. The tiny gametophyte of the fern, which contrasts sharply with the large size of the sporophyte, is typical of all terrestrial vascular plants. The conspicuous generation of all terrestrial vascular plants reproduces by (spores/gametes) ____________________ because it is the (gametophyte/sporophyte) ____________________ generation.

- - - - - - - - - - - - - - - - - - -

spores; sporophyte

39. In view of the information discussed in frames 37 and 38, where would you expect to find the gametophyte generation of the club moss?

______________________________ What would be its relative size?

____

- - - - - - - - - - - - - - - - - - -

We would expect to find the gametophyte growing on the soil, and we would expect it to be relatively tiny.

40. At the beginning of this chapter we investigated examples of alternation of generations in which the gametophytes of mosses and liverworts were the dominant free-living generation on which insignificant sporophytes depended. Subsequent discussions of the Boston fern and club moss developed the concept that the conspicuous generation in

these plants is the sporophyte generation, which is better adapted to terrestrial existence.

Write a summary paragraph comparing the life cycle of a typical moss with that of a fern.

- - - - - - - - - - - - - - - - - -

Your answer should have included the following concepts: (1) Both the moss and the fern possess an alternation of generations in their life cycles. (2) The moss gametophyte is the conspicuous generation with the sporophyte dependent on it. (3) The conspicuous generation in the fern is the sporophyte, while the gametophyte is tiny and inconspicuous.

41. The moss gametophyte represents the maximum expression of a gametophytic land form, but it is limited in size by the lack of conducting (vascular) tissue and the need for a watery medium for its sperms.

Explain the statement, "The extent to which a moss gametophyte can live on land is limited by its structure and function."

____

____

- - - - - - - - - - - - - - - - - -

The gametophyte cannot grow to any great height because it lacks vascular tissue and needs water for the sperm to swim to and fertilize the egg.

42. A comparison between a moss sporophyte and the sporophyte generation of a vascular plant with its large, free-living (independent—not attached to the gametophyte) growth form reveals a few important similarities.

    Place a check mark before the statements listed below which show the similarities between the moss sporophyte and the sporophyte of a vascular plant.

    _______ a. Both possess vascular tissue.

    _______ b. Both reproduce by spores.

    _______ c. Both are formed as the result of spore germination.

    _______ d. Both possess the 2n chromosome number.

    _______ e. Both form reproductive cells by meiosis.

- - - - - - - - - - - - - - - - - - -

b, d, and e are accurate expressions of these similarities

43. The giant redwood and eucalyptus trees reaching a height of nearly 300 feet may represent the maximum expression of size in a sporophytic land form.

    The great height of the sporophyte generation may be explained by the presence, not only of supporting tissue, but of _____________ tissue as well.

- - - - - - - - - - - - - - - - - - -

vascular

44. It is tempting to speculate that the mosses and liverworts were the first plants in geologic time to live out of water, and that the vascular plants evolved from them. However, neither the fossil record in rocks nor a comparison of the forms of the two groups make this assumption likely. Mosses and liverworts are probably not closely related to the vascular plants.

    Check those of the following statements which describe the probable relationship between the mosses and vascular plants.

    _______ a. Mosses were present on the earth before vascular plants.

    _______ b. Mosses and vascular plants arose as separate groups.

_______ c. At one time in the geologic past the moss sporophyte generation lived independently of the gametophyte generation.

_______ d. Probably vascular plants gave rise to the mosses.

_______ e. Probably mosses gave rise to the vascular plants.

- - - - - - - - - - - - - - - - - -

b describes the probable relationship between the mosses and vascular plants. There is little or no evidence to support the other statements.

SELF-TEST
Alternation of Generations—Patterns of Plant Reproduction

Before proceeding to the next chapter, use the following self-test to measure your understanding of the concepts presented in this chapter. All questions should be answered in the light of information presented in Chapter 2 of this book. Answers are given at the end of the quiz.

1. The moss plant differs from vascular plants in that the dominant generation of the moss is the (sporophyte/gametophyte) ______________.

2. What process occurs within a sporangium, and what name is given to the cells formed by this process? ______________________________

3. Indicate whether the following are gametophytic or sporophytic structures.
   a. fern leaf ____________________
   b. true moss "leaf" ____________________
   c. sporophyll ____________________
   d. true moss capsule ____________________
   e. vascular tissue ____________________

4. In what two ways do sporophytes differ from gametophytes? ________

   ____

5. How are sporophytes of mosses and liverworts alike from the standpoint of nutrition? ____

   How are they alike from the standpoint of chromosome number?

   ____

   How are they alike from the standpoint of method of reproduction?

   ____

____

6. In the life cycle of a fern, meiosis occurs in the (sporangium/sporophyll/strobilus) ____________________.

7. Check the true statements about terrestrial plants.

_______ a. The gametophyte is dependent on the sporophyte.

_______ b. Vascular tissue is limited to the sporophyte generation.

_______ c. Sporophylls never occur in the gametophyte generation.

_______ d. Spores germinate to form gametophytes.

8. What is the structural relationship between sporophylls, strobili, and sporangia? ____

9. Into what two parts is the moss sporophyte divided? _____________

____

10. To summarize the life cycle of a moss plant as it relates to alternation of generations, place the following five stages in correct chronological order by letter only. Begin at any place you desire, proceeding in correct order to complete the life cycle.

(A) zygote, (B) spores, (C) sperm and egg, (D) green leafy moss plant, (E) seta and capsule

11. Identify each of the five stages of the moss life cycle listed in question 10 as <u>n</u> (haploid) or 2<u>n</u> (diploid). ___________________________

____

12. If you were to insert the word "meiosis" in the proper place in the moss life cycle stages listed in question 10, it would be placed between the items lettered ______ and _______.

13. Which of the five stages listed in question 10 are part of the sporophyte generation? ______________________________

14. As a summary of your understanding of alternation of generations, use the following words to construct a diagram of alternation of generations for the Boston fern. Use the letters corresponding to the items to make your diagram of the fern life cycle.

    (A) sporophylls, (B) meiosis, (C) fertilization, (D) leafy Boston fern, (E) gametophyte, (F) spores, (G) sporangia, (H) zygote, (I) gametes.

15. List all of the stages of the life cycle of the Boston fern that are diploid. Select your answers from the choices listed in question 14 and answer by letter only. ______________________________

Answers

1. gametophyte (frame 26)

2. Spores are formed by the process of meiosis in a sporangium. (frames 30, 32)

3. a. sporophytic (frame 28)
   b. gametophytic (frame 3)
   c. sporophytic (frame 31)
   d. sporophytic (frame 8)
   e. sporophytic (frame 27)

4. Manner of reproduction (spores vs. gametes) and chromosome number. They also differ in appearance. (frames 11-14)

5. Both sporophytes are dependent on their gametophytes. Both are diploid. Both form spores by the process of meiosis. (frames 7, 18)

6. sporangium (frames 30-32)

7. a. false (frames 38, 39)
   b. true (frames 27, 41)
   c. true (frames 31, 32)
   d. true (frame 37)

8. Sporangia are located on sporophylls; sporophylls are grouped to form strobili. (frames 30, 33, 35, 36)

9. seta and capsule (frame 8)

10.

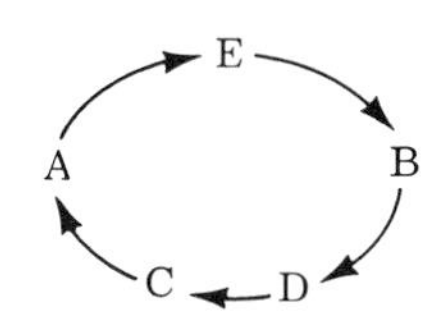

(frames 13-15)

11. A is $2\underline{n}$; B is $\underline{n}$; C is $\underline{n}$; D is $\underline{n}$; E is $2\underline{n}$ (frame 14)

12. E; B (frame 13)

13. A and E (frames 8-15)

14.

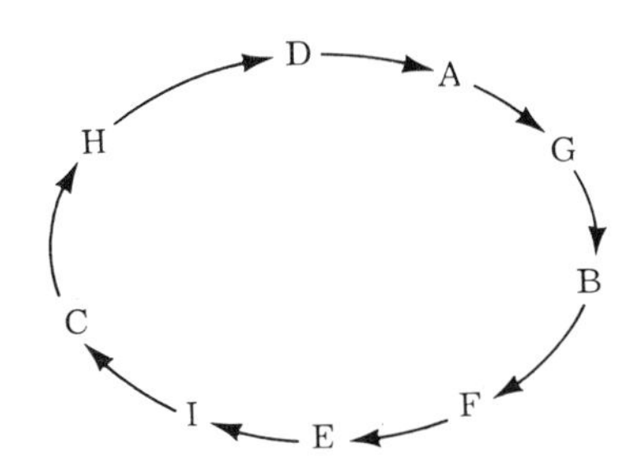

(frames 26-34, 37, 38)

15. A, D, G, H (frames 26-34, 37, 42)

CHAPTER THREE

Heterospory

1. In Chapter 2, you learned that the sporophyte generation bears that name because it reproduces by the formation of (gametes/spores) ____________________ in special containers called ________________.

- - - - - - - - - - - - - - - - - -

spores; sporangia

2. You should recall also that spore production in alternation of generations is accomplished by (mitosis/meiosis) ____________________ to produce (haploid/diploid) ____________________ spores.

- - - - - - - - - - - - - - - - - -

meiosis; haploid

3. The gametophyte generation begins within the diploid spore-producing structures (sporangia) of the sporophyte generation when meiosis gives rise to haploid spores. Recall that the leafy Boston fern is the sporophyte generation and that sporangia are borne on the underside of its sporophylls. Which of the following parts of the Boston fern are diploid (2n) and which are haploid (n)?

_______ a. sporangium

_______ b. fern leaf

_______ c. spore

_______ d. sporophyll

_______ e. gametophytic generation

- - - - - - - - - - - - - - - - - -

a. 2n; b. 2n; c. n; d. 2n; e. n

4. Each of the plants studied in the two previous chapters produces spores of uniform size. The production of equal-sized spores is called homospory (literally, "similar spores") and a plant producing such spores is described as homosporous.

   Is a moss plant homosporous? _______ Why? _______________

   ____

- - - - - - - - - - - - - - - - - -

yes, because its spores are uniform in size

5. In many vascular plants, the sporophyte generation produces two kinds of spores which differ in size. The production of spores of unequal size is called heterospory (literally, "different spores") and the plant is said to be heterosporous.

   A certain plant possesses sporangia containing many small spores. The same plant also produces sporangia containing a few large spores.

   What name can be used to describe spore production in this plant?

   ___________________ Why? ___________________________

   ____

- - - - - - - - - - - - - - - - - -

heterospory, because the plant produces spores of two different sizes

6. The spores of a heterosporous plant typically differ significantly in size. The larger spores, called megaspores (literally, "large spores"), can often be seen with the naked eye while the smaller spores or microspores (literally, "small spores") can be seen only with a microscope.

   A heterosporous plant produces large spores called

   ___________________ and small ones called ___________________.

- - - - - - - - - - - - - - - - - -

megaspores; microspores

7. Megaspores and microspores are often formed by a single sporophyte, but in other instances, a single sporophyte forms only megaspores or microspores. If a single plant is capable of producing both megaspores and microspores, you may logically ask if both kinds of spores can be formed in a single sporangium. They cannot. Megaspores form in special sporangia called megasporangia and microspores occur only in microsporangia.

   A single plant producing two kinds of spores will possess how many kinds of sporangia? ____________________

- - - - - - - - - - - - - - - - - -

two—megasporangia and microsporangia

8. Megaspores are formed in __________sporangia, while microsporangia form only __________spores.

- - - - - - - - - - - - - - - - - -

mega; micro

9. Megaspores are formed in megasporangia by (mitosis/meiosis) ____________________. Microspores are formed by (mitosis/meiosis) ____________________ and will have the (haploid/diploid) ____________________ number of chromosomes.

- - - - - - - - - - - - - - - - - -

meiosis; meiosis; haploid

10. In Chapter 2, you learned that with very few exceptions, sporangia of vascular plants are formed on modified leaflike organs called sporophylls.

    The megasporangia and microsporangia of a heterosporous sporophyte are also found on sporophylls. If a sporophyll which bears a megasporangium is called a megasporophyll, what name is properly applied to the sporophyll bearing a microsporangium? ____________

- - - - - - - - - - - - - - - - - - - -

microsporophyll

11. Recall that an aggregation of sporophylls is called a strobilus or cone. A cone may be composed entirely of megasporophylls, entirely of microsporophylls, or of a combination of both.

    Suppose you had before you a homosporous and a heterosporous plant. In what ways would you expect the composition of the strobili of these two plants to differ? ____________________________

    ____

    ____

    ____

- - - - - - - - - - - - - - - - - - - -

In a homosporous plant all of the sporophylls that bear sporangia would be similar, and all spores produced would be of the same size. In the heterosporous plant a single strobilus could be composed entirely of microsporophylls whose sporangia bear only microspores or entirely of megasporophylls whose sporangia produce only megaspores, or of a combination of both megasporophylls and microsporophylls.

12. Megaspores and microspores, like the spores of a homosporous plant, will germinate to form gametophytes. The prefixes, mega and micro, can be applied to those gametophytes arising from heterospores.

    Thus, megaspores give rise to (mega/micro) __________gametophytes and microspores form (mega/micro) __________gametophytes.

- - - - - - - - - - - - - - - - - - - -

megagametophytes; microgametophytes

13. The gametophytes of a heterosporous plant are different from those formed by a homosporous plant. In homosporous plants the spores germinate and give rise to a gametophyte that is much larger than the spore itself. In heterosporous plants the megagametophytes and

microgametophytes are smaller than the spores and invariably develop inside the walls of the megaspores and microspores, respectively. Later you will see how this applies to the life cycles of several heterosporous plants.

Which of each pair following is the larger structure?

a. microgametophyte or microspore ____________________

b. megaspore or megagametophyte ____________________

c. megaspore or microspore ____________________

d. microgametophyte or megagametophyte ____________________

- - - - - - - - - - - - - - - - - -

a. microspore; b. megaspore; c. megaspore; d. megagametophyte

14. The sexes of the gametophytes of heterosporous plants are invariably fixed. Thus, megagametophytes are always female gametophytes and microgametophytes are male. Therefore, microgametophytes produce (eggs/sperms) ____________________ and megagametophytes produce (eggs/sperms) ____________________.

- - - - - - - - - - - - - - - - - -

sperms; eggs

15. Although the literal meaning of megagametophyte is "large gametophyte," and the meaning of microgametophyte is "small gametophyte," in this context the prefixes mega and micro can be considered synonymous with the terms female and male. As a review of the material dealing with the concept of heterospory, answer the following questions.

a. Based on the type of spores produced, into what two categories can spore-producing plants be divided? ____________________

____

b. Suppose a heterosporous plant produces strobili (cones). What structure(s) may comprise a single strobilus? ______________

____

c. A megagametophyte produces what kind of gametes? __________

d. What structure immediately surrounds a microgametophyte?

____

e. Name two kinds of structures formed by meiosis in a heterosporous plant. ____________________

f. What kind(s) of spore cases can be found on a heterosporous sporophyte? ____________________

g. Indicate whether each of the following is haploid or diploid.

(1) megasporophyll ____________________

(2) microsporangium ____________________

(3) megaspore ____________________

(4) megagametophyte ____________________

(5) sporophyte plant ____________________

- - - - - - - - - - - - - - - - - -

a. homosporous and heterosporous plants
b. microsporophylls possessing microsporangia only, megasporophylls possessing megasporangia only, or both megasporophylls and microsporophylls.
c. eggs (female)
d. microspore wall
e. megaspores, microspores (either order)
f. megasporangia and microsporangia (either order)
g. (1) diploid; (2) diploid; (3) haploid; (4) haploid; (5) diploid

16. Let us now direct our attention to the life cycle of a plant in which heterospory is clearly evident. The club moss, Selaginella, a relative of the ferns (and not a true moss), is such a plant.

Selaginella is a vascular plant having small, delicate stems covered by four longitudinal rows of tiny leaves. The tips of many of these stems are covered with clusters of modified leaves which contain sporangia in which spores are produced. As such, these modified leaves bearing sporangia are properly called ____________________ and the conelike clusters of leaves are technically called ____________.

- - - - - - - - - - - - - - - - - -

sporophylls; strobili

17. This means that the green leafy Selaginella plant is a (gametophyte/ sporophyte) ____________________, and that it is made up of cells which contain the (haploid/diploid) ____________________ chromosome number.

- - - - - - - - - - - - - - - - - - -

sporophyte; diploid

18. Careful examination of a strobilus of Selaginella reveals that it consists of two kinds of sporophylls: those near the tip of the strobilus, which bear microsporangia, and those near the base, which bear megasporangia. The sporophylls near the tip of the strobilus which bear microsporangia are called ____________________ and those near the base bearing megasporangia are called ____________________.

- - - - - - - - - - - - - - - - - - -

microsporophylls; megasporophylls

19. A microsporangium contains microspore mother cells in which (meiosis/mitosis) ____________________ occurs to yield small cells called ____________________.

- - - - - - - - - - - - - - - - - - -

meiosis; microspores

20. In the microsporangium many microspore mother cells divide by meiosis to yield many microspores, while in a megasporangium a single megaspore mother cell divides by meiosis to produce only four ____________________.

- - - - - - - - - - - - - - - - - - -

megaspores

21. Each haploid microspore nucleus divides by the process of mitosis to

retain the haploid chromosome number in the cells of the resulting (microgametophyte/megagametophyte) ____________________. Several cells of the microgametophyte become male gametes or sperms which are enclosed by a protective layer of cells.

- - - - - - - - - - - - - - - - - -

microgametophyte

22. Similarly, each haploid <u>megaspore</u> undergoes mitotic divisions to produce a haploid gametophyte called a/an ____________________. This structure includes food-storing cells and several female gametes or ____________________.

- - - - - - - - - - - - - - - - - -

megagametophyte; eggs

23. To summarize, in <u>Selaginella</u> the large conspicuous club moss plant is the sporophyte generation while the gametophyte generation is inconspicuous, microscopic, and may be completely dependent on the green leafy <u>Selaginella</u> plant for nutrition. How does the life cycle of <u>Selaginella</u> compare and contrast with that of the true moss discussed in Chapter 2?

- - - - - - - - - - - - - - - - - -

Your answer should include the following ideas: (1) Both plants have an alternation of generations. (2) The conspicuous generation of the true moss is the green leafy gametophyte, while the conspicuous <u>Selaginella</u> plant is the sporophyte generation. (3) In the true moss the sporophyte is dependent on the gametophyte, while in <u>Selaginella</u> the gametophyte is dependent on the sporophyte.

24. Microspores, each containing the male gametophyte with its male gametes or sperms, drop from the microsporophylls at the top of the strobilus or cone to the vicinity of the megagametophytes. These megagametophytes may still be retained in the megasporangium on megasporophylls at the base of the strobilus, but more commonly they are discharged from the megasporangia and are found lying on the damp soil. Here in a drop of water the microspores germinate and release the male gametes into the water. The sperms swim to the vicinity of the eggs and unite with them in the megagametophyte. Fertilization is thus accomplished and the diploid chromosome number is restored in the zygotes. Each zygote then undergoes mitotic divisions which result in the formation of an embryo, all the cells of which are diploid. Thus, the zygote is the first cell in the sporophyte generation. Competition among the several embryos for the food stored in the megagametophyte usually results in the survival of a single embryo. The surviving embryo sporophyte thus develops within the megagametophyte. Initially the new sporophyte lacks chlorophyll and cannot carry on photosynthesis, the process of manufacturing its own food. Later it develops chlorophyll and becomes capable of survival as a separate individual because it can photosynthesize.

Study the word diagram of the Selaginella life cycle which summarizes the discussion to this point. Complete this diagram by filling in the missing words.

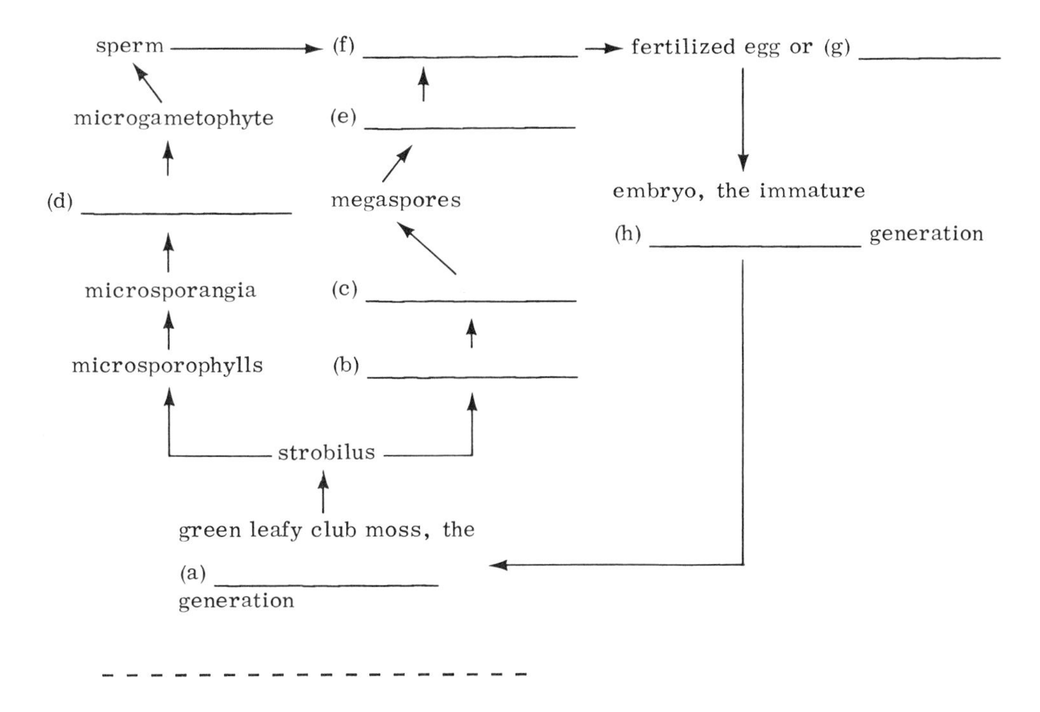

- - - - - - - - - - - - - - - - - - - -

a. sporophyte; b. megasporophylls; c. megasporangia; d. microspores; e. megagametophyte; f. egg; g. zygote; h. sporophyte

25. In the diagram of the Selaginella life cycle shown in frame 24 the word "meiosis" could be inserted between the words ______________ and ____________________ and between ____________________ and ____________________ to indicate where in the life cycle the process of meiosis occurs.

- - - - - - - - - - - - - - - - - -

microsporangia and microspores; megasporangia and megaspores (either order)

26. Where should the word "fertilization" be inserted? ______________
____

- - - - - - - - - - - - - - - - - -

following sperm and egg and prior to fertilized egg or zygote

27. To further summarize the life cycle of Selaginella, place a check mark before each of the following stages that are diploid.

_______ a. microspore

_______ b. megasporophyll

_______ c. megagametophyte

_______ d. zygote

_______ e. strobilus

_______ f. sperm

_______ g. green leafy club moss plant

_______ h. microsporophyll

_______ i. microgametophyte

_______ j. sporophyte

- - - - - - - - - - - - - - - - - -

b, d, e, g, h, and j

28. Suppose you were given a mature Selaginella plant and asked to explain its life cycle in terms of heterospory and alternation of gametophyte and sporophyte generations. Provide as much detail as you can.

- - - - - - - - - - - - - - - - - -

Your answer should have included the following key points: (1) The green leafy Selaginella plant constitutes the heterosporous sporophyte generation whose strobili possess megasporangia and microsporangia. (2) Meiosis occurs in the microsporangia and megasporangia and produces haploid microspores and megaspores, respectively. (3) The microspores give rise to microgametophytes among the cells of which are sperms, and the megaspores give rise to egg-producing megagametophytes. (4) Fertilization restores the diploid chromosome number in the zygote. Mitotic divisions result in the formation of a diploid sporophyte embryo which develops in the megagametophyte.

SELF-TEST
Heterospory

To review how heterospory affects the life cycles of plants, complete the following self-test. All questions should be answered in the light of information presented in the first three chapters of this book. Answers are given at the end of the quiz.

1. Review the basic steps in alternation of generations, beginning with the sporophyte generation and inserting the appropriate words in the following sequence. Choose the words from the lettered list and answer by letter only.

   (A) gametes, (B) zygote, (C) gametophyte, (D) sperm, (E) spores, (F) egg, (G) sporophyte.

   sporophyte → meiosis → _______ → mitosis → _______ → gametes (_______ and _______) → fertilization → _______ → mitosis → _______.

2. Selaginella has the alternation of generations pattern in its life cycle, but meiosis results in the production of both _______________ and _______________ which by the process of mitosis give rise to _______________ and _______________, respectively.

3. In Selaginella microspores are produced in spore cases called _______________ which are located on modified leaves called _______________ which constitute part of a conelike structure called a/an _______________.

4. In heterosporous plants the spores usually differ in size but are alike in that both are produced by the process of _______________, are _______________ in chromosome number, and divide by the process of _______________ to give rise to a haploid _______________ generation.

5. Match numbered items with the most appropriate lettered items. Use numbered items more than once if necessary.

_______ a. microgametophyte	1.	large gametophyte
_______ b. homosporous plant	2.	produces eggs
	3.	microsporophylls
_______ c. structure containing a single megagametophyte	4.	microsporangium
	5.	male gametophyte
_______ d. constituents of a Selaginella strobilus	6.	sperms
	7.	liverwort
	8.	female gametophyte
_______ e. produced by meiosis in Selaginella	9.	megaspore
	10.	megasporophylls
	11.	Selaginella

6. Place the following items in order by size and level of complexity, starting with the largest and most complex. You may answer by letter only.

   (A) microsporangia, (B) strobilus, (C) microspores, (D) sporophyte, (E) microsporophyll.

   ____

7. Since it produces both microspores and megaspores, Selaginella is said to be ____________________.

Answers

1. E; C; D, F (either order); B; G (frames 23, 24)

2. microspores, megaspores (either order); microgametophytes; megagametophytes (frames 6, 12, 21, 22)

3. microsporangia; microsporophylls; strobilus (frames 18, 19)

4. meiosis; haploid; mitosis; gametophyte (frames 5, 6, 12)

5. a. 5 (frame 14)
   b. 7 (frame 4)
   c. 9 (frame 13)
   d. 3 and 10 (frame 18)
   e. 9 (frame 20)

6. D-B-E-A-C (frames 7, 8, 10, 11)

7. heterosporous (frame 5)

CHAPTER FOUR

The Formation of Seeds

1. In the preceding discussion of plants with an alternation of generations in their life cycles, the young sporophyte comes into existence at the moment the egg is fertilized by the sperm. Eventually, the fertilized egg or zygote divides to form a multicellular sporophyte.

   Technically, the zygote is the first cell of the sporophyte generation because it is formed by the process of (meiosis/fertilization) ____________________ and bears the (haploid/diploid) ______________ chromosome number.

   \- - - - - - - - - - - - - - - - - -

   fertilization; diploid

2. The members of most plant groups <u>retain</u> their zygotes within the enclosing parental gametophytes where zygote germination takes place immediately. Consequently, a young multicellular sporophyte begins its existence within the gametophyte and obtains its food supply from the gametophyte. A young sporophyte developing under these conditions is called an <u>embryo</u>.

   Which of the following statements accurately describe alternation of generations in most plant groups?

   _______ a. Fertilization may occur outside the parent gametophytes.

   _______ b. Young sporophytes are dependent upon the gametophytes.

   _______ c. The multicellular sporophyte develops within the confining parent gametophyte.

   \- - - - - - - - - - - - - - - - - -

   b and c are correct

3. When alternation of generations occurs in the algae and fungi, the zygote is often formed outside of the parent gametophyte, and in all cases the multicellular sporophyte is not intimately associated with the parent gametophyte.

   Comment on the accuracy of the following statement: Algae and fungi do not possess embryos. ______________________________

   ____

   ____

   - - - - - - - - - - - - - - - - - -

   The statement is true because the zygotes of algae and fungi do not germinate and develop into a sporophyte within the parental gametophyte.

4. When you recall that the gametophytes of all vascular plants are small, poorly developed structures, it will be obvious to you that the embryos of vascular plants develop under precarious conditions. The small gametophytes are unable to furnish their embryos with much food, and they offer the embryo virtually no protection against drought and extremes in temperature. Not surprisingly, few embryos develop into mature sporophytes. The well-known statement, "Many are called but few are chosen," can be applied to embryo development in these plants.

   From the preceding paragraph one can conclude that two major factors contributing to embryo mortality are:

   (1) ____

   (2) ____

   - - - - - - - - - - - - - - - - - -

   (1) lack of food; (2) lack of protection from drought and extremes in temperature (either order)

5. Any plant possessing adaptations which afford its embryos greater protection against a hostile environment will reproduce more efficiently and have a much greater chance for survival. Actually, a large number of vascular plants have evolved structures called seeds which favor embryo development. A seed is a complex device which affords an embryo the needed protection that greatly increases its chances of survival. The seed and its enclosed embryo may remain

dormant through an extended period that is unfavorable for growth of a young plant.

In almost all natural plant populations seed-bearing plants greatly outnumber all other vascular plants. Which of these statements probably explains this phenomenon?

_______ a. Seed plants produce a greater number of spores than other vascular plants.

_______ b. The vascular system of seed plants is better adapted for a terrestrial existence than that of most other plants.

_______ c. A greater number of embryos of seed plants survive than do those of other vascular plants.

_______ d. A greater number of gametophytes of seed-bearing plants survive than do those of other vascular plants.

- - - - - - - - - - - - - - - - - -

c is the only correct answer. You would have no basis for answering a, b, or d and in fact all are untrue. Survival of the seed embryos is is an important reason for seed-bearing plants outnumbering non-seed-bearing vascular plants.

6. A seed enhances the embryo's chance for survival by forming protective coats of tissue around the embryo to prevent desiccation (severe loss of water) and freezing. In addition, the seed provides an adequate supply of stored food to nourish the young sporophyte until it is large enough to manufacture its own food.

   Can you suggest a reason why it is possible for man to use certain seeds as a source of food? ______________________________

   ____

- - - - - - - - - - - - - - - - - -

A seed possesses stored food which is ordinarily used by the young sporophyte in its early development. Man eats both the stored food and the embryo!

7. One characteristic shared by all mature seeds is a low moisture content. We may encounter seeds which exude oil (stored food) when crushed or dissected, but never watery seeds. The relative dryness of seeds is another protective device for embryos. A low water content in the tissues surrounding the embryo, together with membranes

impervious to oxygen, force the embryos of most seeds into a period of dormancy. Hence, during dry or cold periods when it would be fatal for a young plant to try to establish itself, the embryo is kept inert.

Two ways in which most seeds force the embryo into dormancy are by depriving the embryo of ____________________ and ____________________.

\- - - - - - - - - - - - - - - - - - -

oxygen; water (either order)

8. The dormant embryo within the seed resumes its growth—that is, it germinates—when the embryo is provided with an adequate water supply, an adequate supply of oxygen (which occurs naturally by diffusion when the structures enclosing the embryo are wetted), and a favorable temperature.

   Three factors necessary to break the dormancy of most seed embryos are ____ ____.

   \- - - - - - - - - - - - - - - - - - -

   water, oxygen, and proper temperature (any order)

9. During a dry, unseasonably cold spring you plant a garden, but you find that seeds germinate very poorly. How could you explain your poor results? Do you lack the necessary "green thumb" or could you explain the poor germination in terms of environmental conditions?

   ____

   ____

   ____

   \- - - - - - - - - - - - - - - - - - -

   Probably two factors necessary for breaking dormancy and causing seed germination are missing: proper temperature and adequate water supply. The dry cold spring would not affect the oxygen supply needed for seed germination, but the availability of the oxygen to the embryo would be limited because of the dry condition of the seed.

10. Although the embryos of most seeds are forced into dormancy by the lack of water and oxygen, a few seeds, such as those produced by some desert plants, contain a chemical substance which inhibits embryo germination. A large amount of water must enter the seed to wash away the inhibiting substance before the embryo will germinate. Thus, the embryos of such desert plants are prevented from germinating after a brief unexpected shower during the dry season. A few plants possess seeds which will germinate only after the seeds have been exposed to light.

    Two reasons why seeds may delay germination even after being provided with water, oxygen, and proper temperature are __________ __________________________ and ___________________________________.

- - - - - - - - - - - - - - - - - - -

presence of chemical inhibitors; no exposure to light (in either order)

11. We have discussed how a seed provides protection for its embryo; now let us consider how a seed plant may have evolved from a non-seed-bearing ancestor.

    Most botanists believe that heterospory has led to the seed condition by certain modifications of the structures involved in heterospory. Because the megagametophyte produces (eggs/sperms) ______________________, and following fertilization the resulting embryo receives its food and protection from the (megagametophyte/microgametophyte) ______________________, the major modifications of seed development are thought to involve the megagametophyte and its associated sporophytic structures.

- - - - - - - - - - - - - - - - - - -

eggs; megagametophyte

12. How might this come about? Let's see. In addition to possessing heterospory, a seed-bearing plant possesses only <u>one</u> functional megaspore in a megasporangium. This single megaspore must remain permanently within the enveloping megasporangium.

    On the basis of this discussion give two possible reasons why <u>Selaginella</u>, the heterosporous plant to which you were introduced in Chapter 3, does not produce seeds. ______________________________

____

- - - - - - - - - - - - - - - - - -

First, although Selaginella possesses heterospory (as do all seed plants), it produces more than one megaspore in each megasporangium. Second, the megaspores fall out of the megasporangium.

13. In seed plants the single megaspore, remaining enclosed by the megasporangium, eventually produces a megagametophyte. Therefore, the total number of megagametophytes found in a single megasporangium of a seed plant is ____________________.

- - - - - - - - - - - - - - - - - -

one

14. In heterosporous plants, the gametophytes are smaller than the spores which form them and the gametophytes always occur within the spore walls.

    What structure immediately surrounds the megagametophyte of a seed plant? ____________________

- - - - - - - - - - - - - - - - - -

megaspore wall

15. What structure immediately surrounds the megaspore wall of a seed plant? ____________________

- - - - - - - - - - - - - - - - - -

megasporangium

16. The following is a diagrammatic representation of a megagametophyte of a seed plant. Label the structures which enclose the megagametophyte.

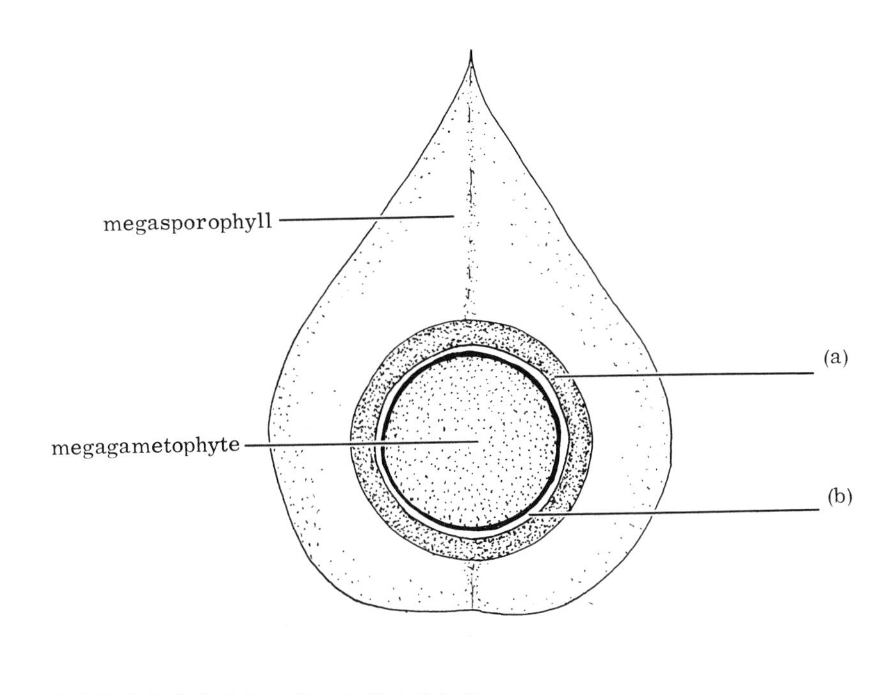

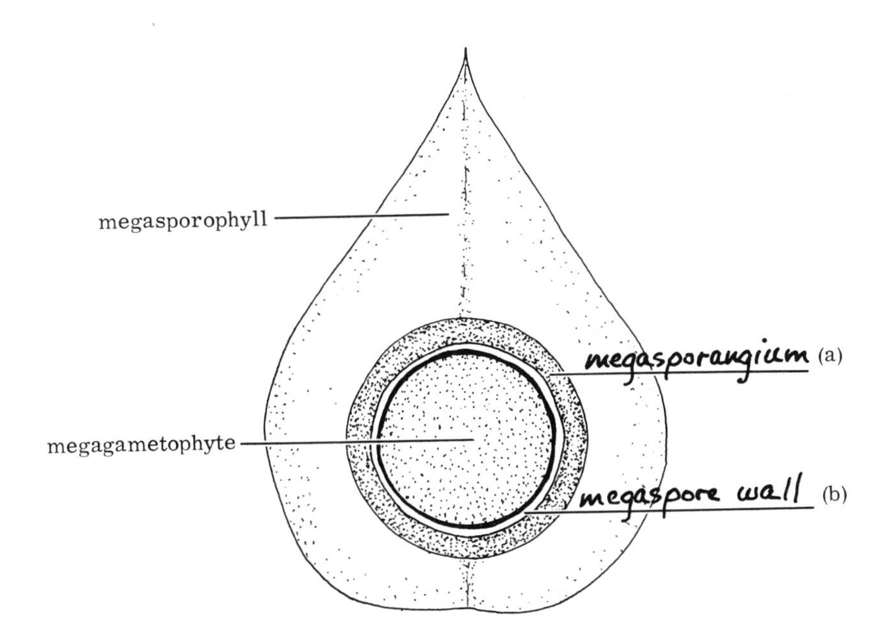

17. The megaspore wall which is represented so prominently in the preceding drawing has been purposely exaggerated in size to emphasize

its position. In seed plants the megaspore wall is actually a thin membranous structure which is soon destroyed and is not recognizable in a dissected megasporangium.

Because of the thin and temporary nature of the megaspore wall, the megagametophyte will appear to be immediately enclosed by the (megaspore/megasporangium) ____________________.

- - - - - - - - - - - - - - - - - -

megasporangium

18. We have indicated that a seed plant possesses (1) heterospory, (2) a single megaspore in each megasporangium, and (3) a permanent retention of that megaspore within the megasporangial wall.

To these three characteristics of a seed plant must be added a fourth one: the presence of one or more layers of protective tissues (the <u>integuments</u>) which almost completely surround the megasporangium.

The megasporangium of a seed plant is not readily visible because it is enclosed by layers of tissues called ____________________.

- - - - - - - - - - - - - - - - - -

integuments

19. However, the megasporangium is not <u>completely</u> surrounded by the enclosing integuments because a microscopic pore-sized opening is left by the failure of the integuments to close. This small pore is called a <u>micropyle</u> (literally, "tiny pore").

In the following diagram label the micropyle and the enclosing structure around a young megasporangium.

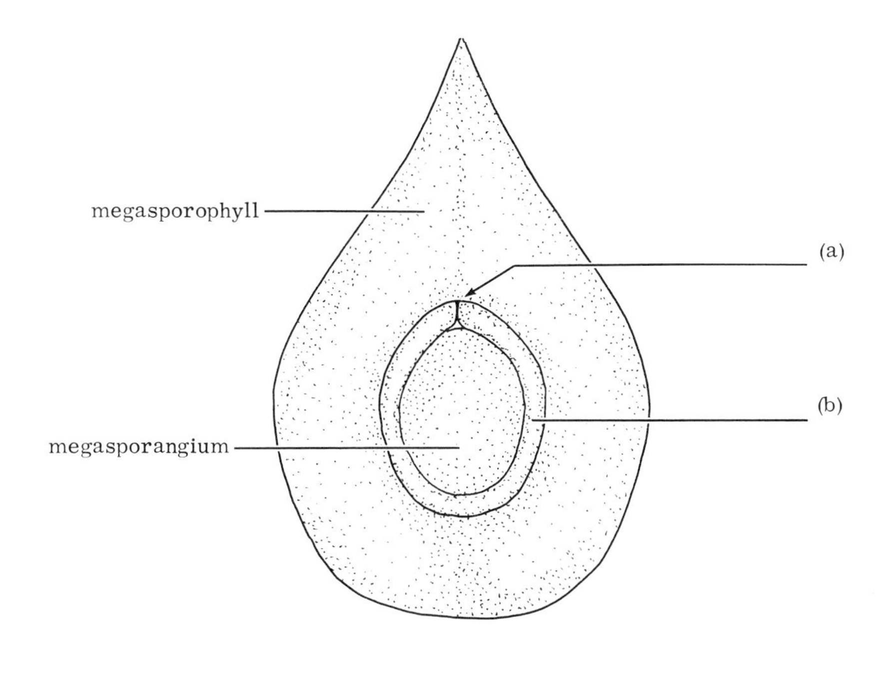

- - - - - - - - - - - - - - - - - -

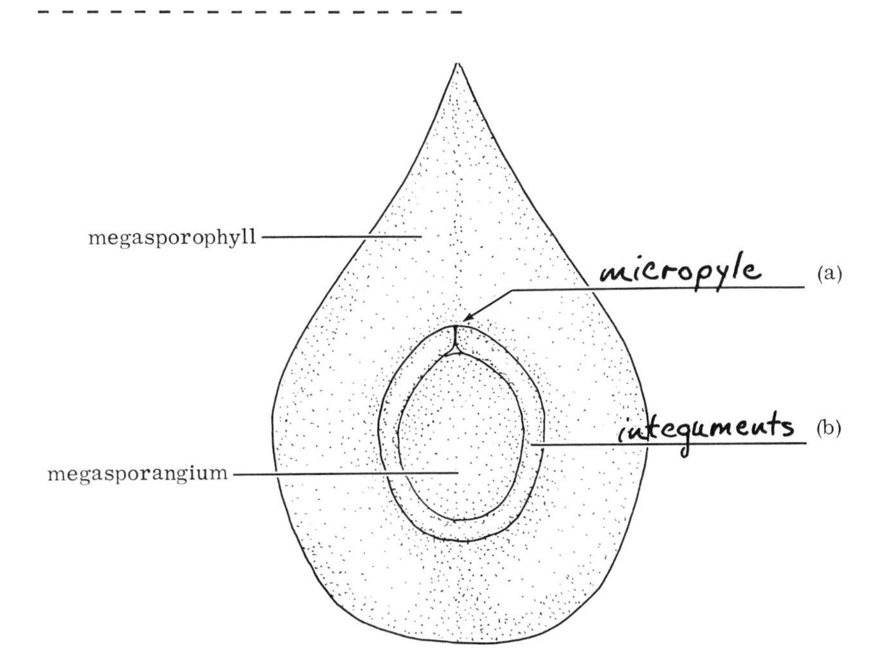

20. Through long usage, the term ovule is applied to the structure composed of the megasporangium plus its surrounding integuments. The

term is inappropriate, for ovule literally means a little egg. It is a convenient word, however, which we will use throughout the rest of this textbook.

An ovule is composed of a ____________________ which is surrounded by ____________________.

- - - - - - - - - - - - - - - - - - -

megasporangium; integuments

21. Many of the structures involved in seed formation were named before their true heterosporous nature was understood. The term ovule is only one example. The megasporangium is more often called the nucellus (literally, "little nut"), a term we will use frequently. Keep in mind, however, the true nature of the nucellus; it is never incorrect to call it by its true name—megasporangium.

Place a check before the correct definitions.

_______ a. Ovule—a megasporangium and integuments.

_______ b. Ovule—a megagametophyte and megasporangium.

_______ c. Ovule—a nucellus plus integuments.

_______ d. Nucellus—another name for a megagametophyte.

- - - - - - - - - - - - - - - - - - -

a and c are correct

22. Since a megagametophyte which has developed from the functional megaspore within the megasporangium is (male/female) ____________________, the megagametophyte in a seed plant will produce (sperms/eggs) ____________________.

- - - - - - - - - - - - - - - - - - -

female; eggs

23. The eggs produced by the megagametophyte are formed in the area closest to the micropyle (see diagram).

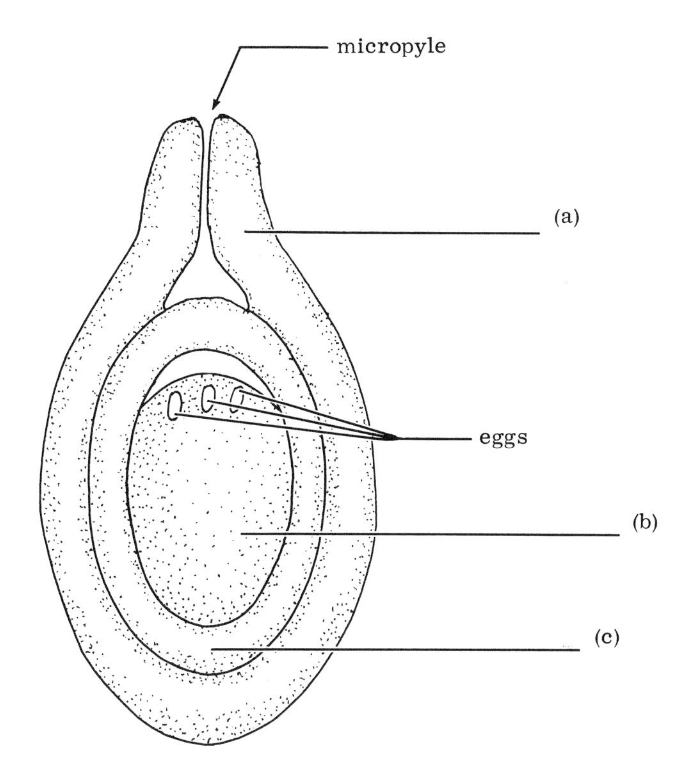

Label the nucellus, megagametophyte, and integuments in the diagram.

- - - - - - - - - - - - - - - - - - - -

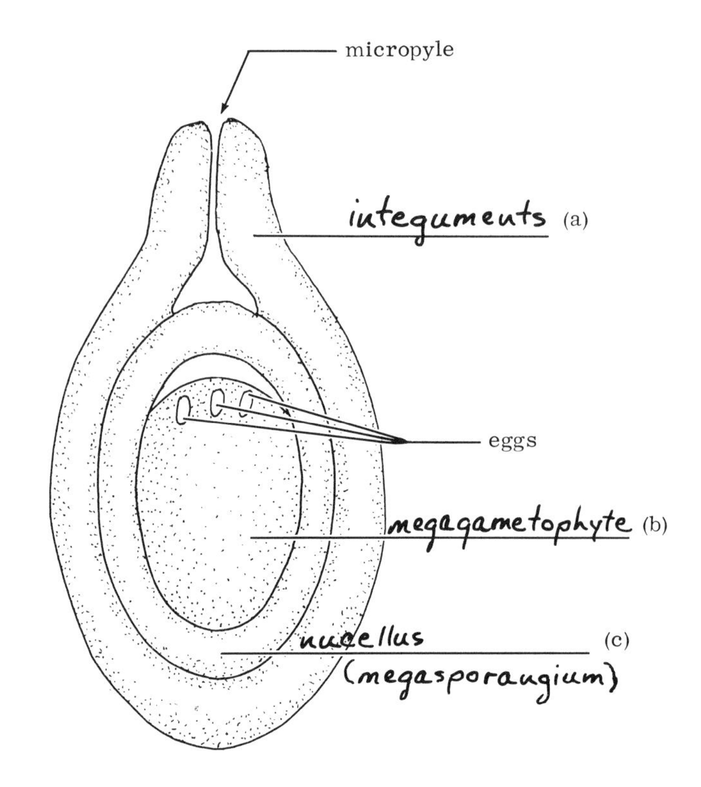

24. Now to review terminology, label the following diagram by choosing your answer from among the numbered choices.

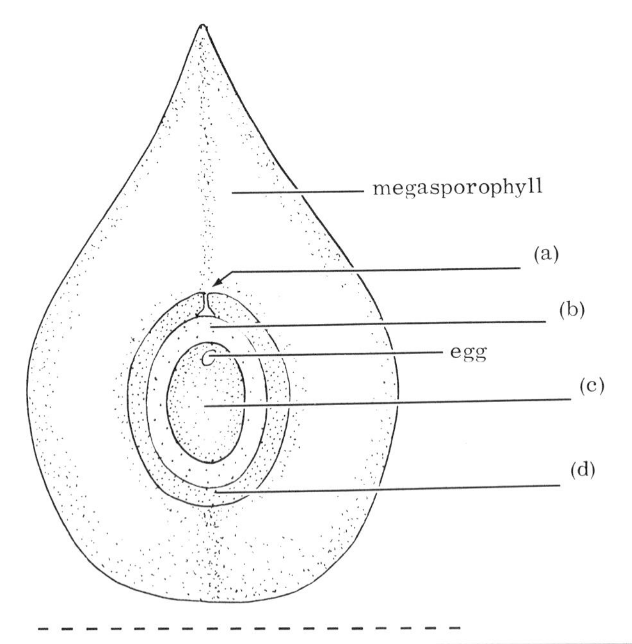

1. nucellus
2. megaspore
3. integuments
4. micropyle
5. megagametophyte
6. megasporangium

- -

a. 4; b. 1 or 6; c. 5; d. 3

25. To further clarify the terminology problems and your understanding of the plant parts represented by these terms, refer to the drawing in frame 24 to help you place the following items in sequence in the order in which they are produced.

(A) megaspore, (B) megasporophyll, (C) egg, (D) megagametophyte, (E) megasporangium

____

\- - - - - - - - - - - - - - - - - -

B-E-A-D-C

26. Check the statements below that are true.

_______ a. Seed plants are generally homosporous.

_______ b. Seed plants permanently retain the functional megaspore within their megasporangia.

_______ c. Seed plants characteristically produce several functional megaspores in each megasporangium.

_______ d. Ovule—megasporangium plus integuments.

\- - - - - - - - - - - - - - - - - -

b and d are true

27. Most of the discussion in this chapter has dealt with the development of the megagametophyte and with the importance of the megagametophyte in seed formation. Now we will examine microgametophyte production in seed plants.

Microsporangia and microspore formation in a seed-producing plant differs very little from that described for Selaginella. Varying numbers of microsporangia are produced on microsporophylls. Inside the microsporangia are produced large numbers of (megaspores/microspores) ____________________ by (mitosis/meiosis) ____________________.

\- - - - - - - - - - - - - - - - - -

microspores; meiosis

28. Microspores develop into microgametophytes which are liberated as the walls of the microsporangia split open. The microgametophytes of seed plants are called pollen or pollen grains (literally, "fine flour" or "dust").

    Check the true statements listed below.

    _______ a. A pollen grain is another name for microspore.

    _______ b. A pollen grain is technically an individual plant.

    _______ c. Microspores form pollen grains.

    _______ d. Pollen and microgametophytes are interchangeable terms.

- - - - - - - - - - - - - - - - - - -

b, c, and d are true statements. b is true since the pollen grain is the microgametophyte and as such is an individual plant.

29. Microgametophytes are (male/female) ____________________ gametophytes and can be expected to produce (sperms/eggs) ______________.

- - - - - - - - - - - - - - - - - - -

male; sperms

30. It is impractical for sperms to swim to the eggs in a terrestrial seed plant. Therefore, any evolutionary modification which allows the microgametophyte with its sperms to be transferred to the vicinity of the egg-producing megagametophyte greatly increases the chance of the sperms uniting with the eggs. The transfer of the male gametophyte or pollen by wind, insects, or other agents to the vicinity of the megagametophyte enclosed within the megasporangium of the ovule is called pollination.

    In Chapter 3 we learned that vascular tissue constitutes an adaptation to terrestrial existence. In the following space discuss why many botanists regard pollination as an adaptation to terrestrial life.

- - - - - - - - - - - - - - - - - - - -

Your answer should include the following ideas: Since sperms cannot swim to eggs in a terrestrial environment, the transfer of the entire sperm-containing microgametophyte to the vicinity of the egg-producing megagametophyte is a necessary prelude to fertilization.

31. Because the small opening or micropyle penetrates the integuments of the ovule, pollination in many seed plants results in some of the pollen or microgametophytes passing through the micropyle and coming to rest on the surface of the megasporangium or **nucellus.** Thus, the only tissue which separates these pollen grains from the megagametophyte after pollination has occurred is that of the nucellus.

    Shortly after coming to rest on the megasporangium or nucellus, the pollen grain sends out a long slender pollen tube which completely penetrates the nucellus. Study and label the following diagram which illustrates the above description.

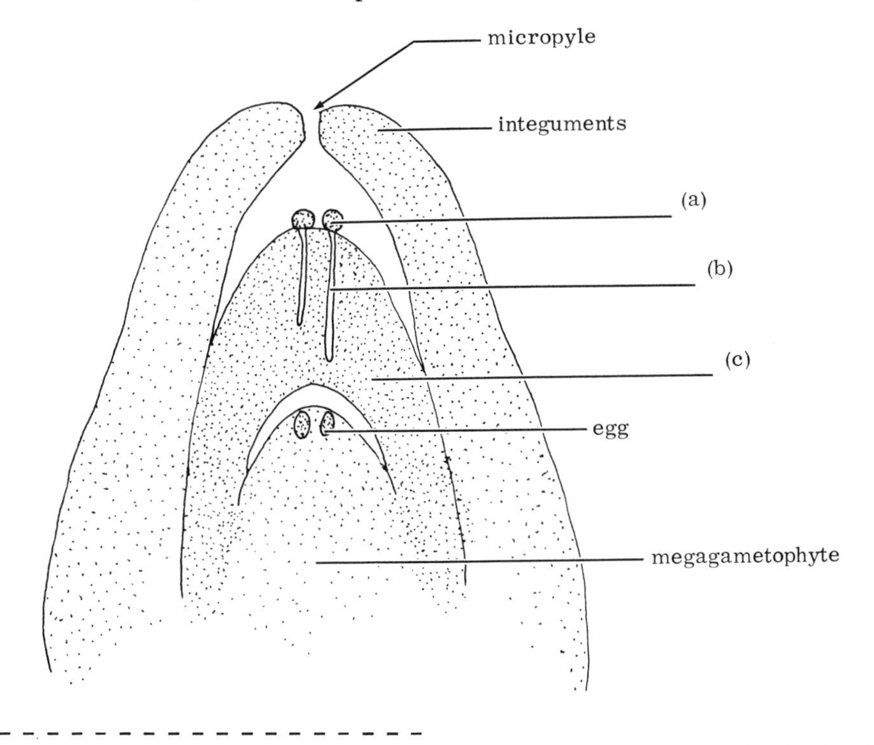

- - - - - - - - - - - - - - - - - - - -

a. pollen grain or microgametophyte; b. pollen tube; c. nucellus or megasporangium

32. Since the pollen grain is a microgametophyte, you should expect it to form male gametes or sperms. This, in fact, happens. As the pollen tube grows through the nucellus one cell within the pollen tube divides to form two sperms. These sperms are discharged near the egg or eggs in the micropylar end of the megagametophyte as the pollen tube ruptures once it penetrates the nucellus. Fertilization quickly follows and one or more fertilized eggs or zygotes are formed within the megagametophyte.

    To summarize the ideas in frames 31 and 32, complete the following matching exercise.

_______ a. The pollen tube grows through the _____ to reach the megagametophyte.

_______ b. The eggs are immediately surrounded by the _____.

_______ c. The pollen grain is the same as the _____.

_______ d. The pollen tube finally contains two _____.

_______ e. The pollen gains access to the nucellus through a pore called the _____ which is located in the integuments.

1. eggs
2. megagametophyte
3. integuments
4. nucellus
5. ovule
6. microgametophyte
7. micropyle
8. sperms
9. seed
10. zygote

- - - - - - - - - - - - - - - - - -

a. 4; b. 2; c. 6; d. 8; e. 7

33. You should note that fertilization occurs after pollination has occurred and as a result of pollination. Thus, fertilization and pollination are not interchangeable terms.

    Suppose 35 pollen grains are lying on the surface of a nucellus of a particular ovule.

a. How many pollen tubes could grow through the nucellus of this ovule? _______

b. Theoretically how many eggs could be fertilized in this ovule providing that sperms can move from egg to egg until an unfertilized one is reached? ________

- - - - - - - - - - - - - - - - - -

a. 35 pollen tubes
b. 70 eggs, because each pollen grain produces two sperms

34. The number of eggs produced by the megagametophyte in an ovule varies with the particular kind of seed plant; most seed plants produce a single egg in each ovule.

    If an ovule possessing a single egg is pollinated by 35 pollen grains which germinate pollen tubes and form sperms in the normal manner, how many pollen grains will be needed to effect fertilization of the egg? ________________

- - - - - - - - - - - - - - - - - -

One pollen grain is needed to fertilize the egg. Actually, one pollen grain produces two sperms, only one of which, theoretically, is needed for fertilization.

35. Until now we have not referred to the ovule as a seed, but with the fertilization of the egg and with the establishment of an embryo within the the ovule, the term seed can now be used. In other words, a seed may be defined as an ovule containing an embryo. Because many factors prevent pollination and/or fertilization of the egg within an ovule, there may be considerably fewer seeds produced than there are ovules.

    The commercial banana is unable to produce normally functioning pollen and ovules because each cell of the sporophyte contains three sets of chromosomes (3*n* or triploid) which interfere with meiosis at the time of microspore and megaspore formation. Consequently the banana does not form seeds.

    The banana fruit is a potential seed-producing structure. What do you think the tiny black specks in the center of the banana are?

____

- - - - - - - - - - - - - - - - - -

The small black bodies are ovules whose eggs have not been fertilized.

36. The young seed embryo requires a large amount of energy as it increases in size during early growth, and the food stored within the seed supplies that energy. The abundant stored food in the form of starch, oils, and proteins is used by man for food and for various commercial purposes. Castor oil, for example, is obtained by extraction of the oil from the seeds of castor bean, and corn starch is obtained from processing corn seeds. Can you suggest why these products (oil and starch) should be highly concentrated in seeds?

____

____

- - - - - - - - - - - - - - - - - - -

The oil and starch are forms of stored food which are available to the embryos of the seeds concerned.

37. Considerable changes occur within a seed as the embryo develops within the tissues forming the ovule. For example, additional foods may be stored in the seed, and the natural enlargement of the embryo may cause the seed to increase greatly in size. The nucellus or megasporangium becomes absorbed or squeezed out of existence during embryo development while the enlarging integuments change to form hard, dry seed coats.

The following diagram represents a seed. Label the structures which enclose the embryo.

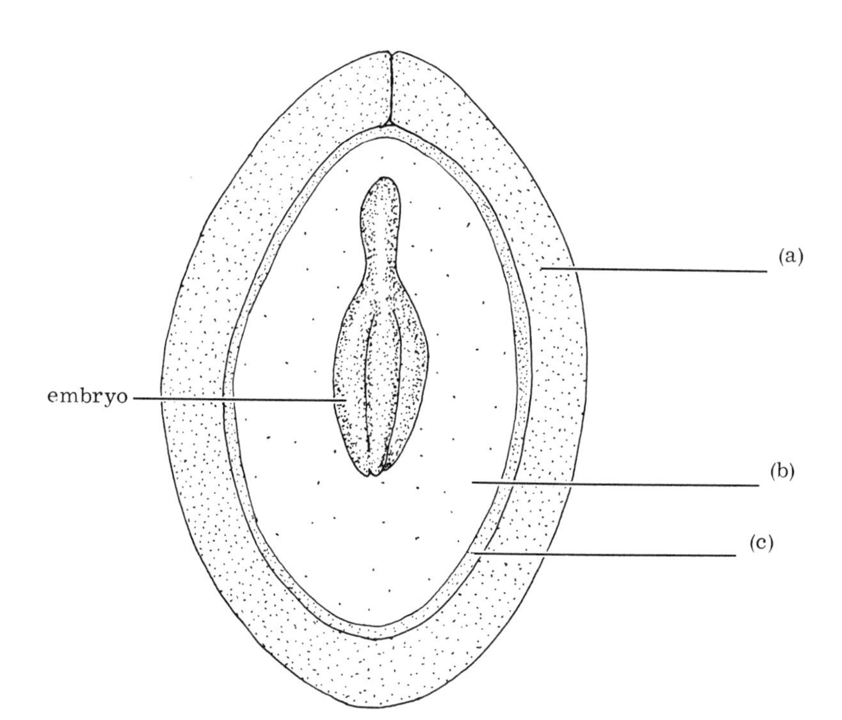

- - - - - - - - - - - - - - - - - -

seed coats (a)

embryo

megagametophyte (b)

nucellus (c)

38. Note in the diagram in frame 37 that the nucellus constitutes only a small part of the young seed. What part of the developing seed would appear to be best suited for food storage? ______________________

____

- - - - - - - - - - - - - - - - - - -

The megagametophyte is nearest to the embryo and provides the embryo with needed food.

39. With the final development of the seed and the formation of the seed coats, the seed embryo which grew so rapidly in its early stages now reaches maturity and settles into the dormancy described in frames 7-10 of this chapter. With proper storage of the seeds, the dormant embryo may remain alive for many years and be ready to resume growth when conditions are favorable.

The embryos of some seeds do not resume growth even when the seeds are provided with adequate moisture, temperature, and oxygen because the seed coats are impermeable to water and oxygen. To assure prompt germination the seed coat must be scarified (abraded or a portion removed). Thus, when the seed coats are nicked with a razor blade, the seeds germinate readily. Which of the following statements might provide an explanation for the behavior of the nicked seeds?

_______ a. Nicking the seed coats provides a mechanical stimulus needed to break the dormancy of the seed embryos.

_______ b. The seed coats are too hard and waxy to allow water or oxygen to diffuse into the seed.

_______ c. Nicking the seed coats provides the embryo with needed minerals from the soil which can now penetrate the seed.

- - - - - - - - - - - - - - - - - - -

b is correct

40. To summarize your understanding of how seeds are produced, use the following words to write a brief paragraph describing seed formation: megaspores, megasporangia (nucellus), megagametophyte, egg, microspores, microsporangia, microgametophyte (pollen), pollen tube, sperm, fertilization, ovule, seed.

____

- - - - - - - - - - - - - - - - - - -

Your answer should include the following points: The megaspores are produced in the megasporangium or nucellus of the ovule. One megaspore forms a single egg-producing megagametophyte within the megasporangium. Similarly, microspores are formed in microsporangia. The microgametophytes (pollen) formed from the microspores germinate pollen tubes on the surface of the megasporangium. The sperms that form in the pollen tube are released in the vicinity of the eggs. Fertilization follows and the embryo develops from the zygote. The ovule which then contains an embryo is now called a seed.

SELF-TEST
The Formation of Seeds

Before you study the details of the life cycles of several different seed plants, check your learning progress concerning the process of seed development by taking this short self-test. All questions should be answered in the light of information presented in this book. Answers are given at the end of the test.

1. Which of the following statements are true? (Place a check before them.)

   _______ a. Pollination always precedes fertilization in the seed plants.

   _______ b. Seed plants are always heterosporous.

   _______ c. In seed plants the megaspore is retained and develops into a megagametophyte within the megasporangium.

   _______ d. A seed may be defined as an embryo surrounded by megasporangium and integuments.

   _______ e. Although water must be present for sperms to reach the egg in a moss plant, water is not necessary for fertilization in seed plants.

2. Over the years that sexual reproduction of seed plants has been studied some terminology has been duplicated. In the following exercise match the numbered items with their equivalent lettered item.

   _______ a. Pollen grain = _____.

   _______ b. Megasporangium plus integuments = _____.

   _______ c. Megasporangium is the same as _____.

   _______ d. Female gametophyte = _____.

   _______ e. Transfer of the male gametophyte to the vicinity of the female gametophyte is called _____.

   1. seed
   2. megagametophyte
   3. microspore
   4. microsporophyll
   5. nucellus
   6. pollination
   7. microgametophyte
   8. integuments
   9. ovule
   10. fertilization
   11. egg
   12. correct answer not given

_______ f. A/an _____ is an ovule containing an embryo.

_______ g. The hard, dry, protective seed coats are derived from the _____.

3. Each of the following questions must be answered by giving a number.

   a. Like the banana, the tomato fruit is a potential seed-producing structure. Suppose you slice a tomato and find that it contains 78 seeds. What is the minimum number of ovules that were produced in the tomato? _______

   b. How many sperms had to function to produce the 78 seeds in the tomato? _______

   c. Since tomato seeds generally contain just one embryo, it would be safe to conclude that _______ eggs were fertilized in producing the 78 seeds.

   d. In seed plants there are typically how many megaspores produced in each megasporangium? _______

4. From what environmental threats does a seed protect its enclosed embryo? ____

5. Name two environmental factors which will maintain the dormancy of a seed. ____

6. What evolutionary advantage is gained by a plant that produces seeds?

   ____

7. Place the following items in sequence in the order in which they are produced.

   (A) microgametophyte, (B) microsporangium, (C) sperm, (D) microsporophyll, (E) microspore

   ____

8. Place the following events in seed plant reproduction in correct chronological order.

   (A) pollination, (B) mature seed germinates, (C) microspore and megaspore formation by meiosis, (D) embryo development from the zygote, (E) male and female gametophyte formation, (F) fertilization

   ____

9. Label the parts of the following diagram by choosing the labels from the list of numbered items.

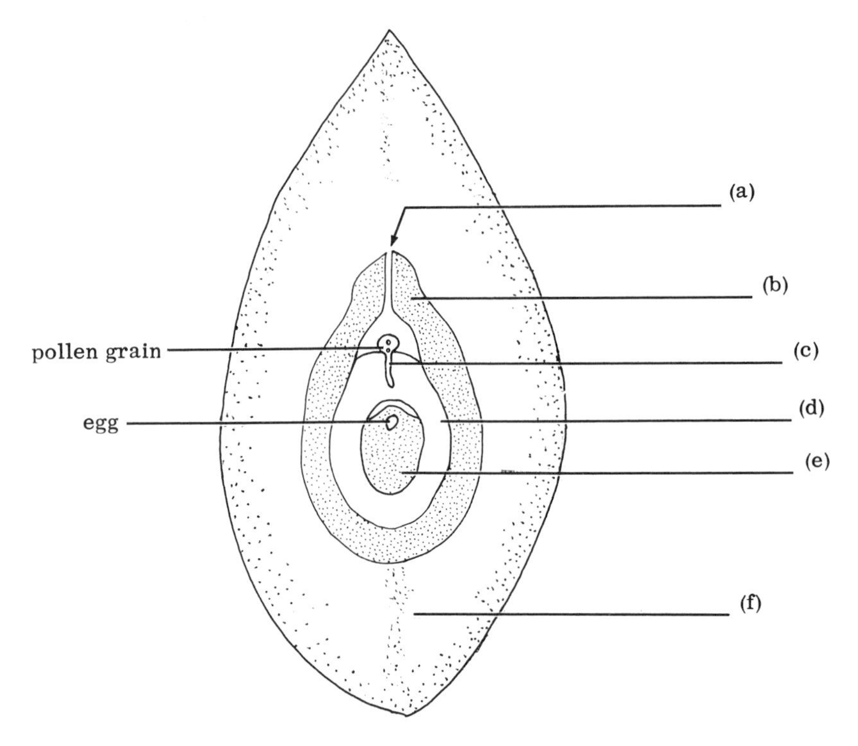

1. microsporophyll
2. integuments
3. megasporangium
4. micropyle
5. ovule
6. megasporophyll
7. nucellus
8. microspore
9. pollen tube
10. microsporangium
11. megagametophyte
12. seed
13. correct answer not given

To summarize the life cycle of a seed plant as we have outlined it in Chapter 4, complete the following questions.

10. A seed plant constitutes the (gametophyte/sporophyte) ____________ generation and as such produces (gametes/spores) ______________ by the process of meiosis. The products of meiosis are produced in structures called ____________________ which are borne on modified leaves called ____________________.

11. The products of meiosis develop into either ____________________, each of which contains two sperm cells, or into ________________ which contain egg cells. Pollination and fertilization follow and the resulting ____________________ develops into an embryo which is enclosed in the seed. The seed is covered by protective seed coats derived from the ____________________. The seed germinates, grows and develops, finally producing a new mature seed plant to complete the life cycle.

Answers

1. All five statements are true. (a. frames 30–32; b. frames 11, 12; c. frames 12, 13; d. frames 18, 35; e. frame 30)

2. a. 7 (frame 28)
   b. 9 (frame 21)
   c. 5 (frame 21)
   d. 2 (frame 22)
   e. 6 (frame 30)
   f. 1 (frame 35)
   g. 8 (frame 37)

3. a. 78 (frame 35)
   b. 78 (frame 32)
   c. 78 (frames 34, 35)
   d. 1 (frame 12)

4. desiccation and freezing (or equivalent answer) (frame 6)

5. low water content and low oxygen availability (frames 7 and 8)

6. Since the seed protects its enclosed embryo, the seed increases the chances of survival for the plant species. (frame 5)

7. (D) microsporophyll, (B) microsporangium, (E) microspore, (A) microgametophyte, (C) sperm. (frames 27, 28, 29, 32)

8. (C) microspore and megaspore formation by meiosis, (E) male and female gametophyte formation, (A) pollination, (F) fertilization, (D) embryo development from the zygote, (B) mature seed germinates. (frames 11–35, 40)

9. a. 4 (frame 19)
   b. 2 (frames 18, 19)
   c. 9 (frames 31, 32)
   d. 3 or 7 (frame 21)
   e. 11 (frame 13)
   f. 6 (frame 16)

10. sporophyte; spores; sporangia; sporophylls (frames 2, 11, 16)

11. microgametophytes; megagametophytes; zygote; integuments (frames 20, 30–33, 37)

CHAPTER FIVE

Kinds of Seed Plants: Gymnosperms

1. In Chapter 4 we discussed the nature of seed plants and the significance of seed production. Let us now consider the various kinds of seed plants, for they are the most numerous and familiar of all of the plant groups in existence today. The great number and variety of seed plants suggests that seed plants are well adapted to life on land. The fossil record reveals that seed-producing plants have been on earth for at least 350 million years.

   Which of the following do you think are true statements about seed plants?

   _______ a. Since coal deposits were formed an estimated 300 million years ago, one may expect to find fossil seed plants embedded in coal.

   _______ b. The abundance of seed plants today indicates their successful adaptation to contemporary environmental conditions.

   _______ c. If you were asked to name the twenty-five plants most familiar to you, in all likelihood you would name mostly seed plants.

- - - - - - - - - - - - - - - - - - -

All three statements are true.

2. Although a variety of seed plants were in existence 350 million years ago, they all possessed one unifying characteristic: their seeds were exposed on the surface of the megasporophylls. No plants then in existence produced seeds enclosed in structures such as the tomato fruit. Naked-seed plants are collectively called <u>gymnosperms</u> (literally, naked seeds).

If you were to examine a seed of a gymnosperm, you would expect to find it (enclosed in a fruit/exposed on a megasporophyll)

________________________ because the word "gymnosperm" implies that the seeds are ______________.

- - - - - - - - - - - - - - - - - - -

exposed on a megasporophyll; naked or uncovered

3. Almost all of the earliest gymnosperms became extinct millions of years ago, but some gymnosperm lines exist today in decreased numbers. Although differing greatly in appearance from their ancestral forms, they are still recognizable as gymnosperms.

   What characteristic do modern gymnosperms share with ancient forms? ________________________

- - - - - - - - - - - - - - - - - - -

the presence of naked or uncovered seeds

4. There exists in many tropical and subtropical areas of the world a small group of tree-sized gymnosperms known as the cycads. Since the cycads are not generally known to most non-botanists, they are often confused with palm trees, which they superficially resemble and to which they are only remotely related.

   Despite their obscurity, the cycads are excellent plants to study because they are thought to possess characteristics similar to those of early gymnosperms that are now extinct. Indeed, many botanists refer to the cycads as living fossils.

   Which of the following statements about cycads would you expect to be true?

   _______ a. Cycads are a kind of palm tree.

   _______ b. Cycads can be expected to produce seeds which are not enclosed in a fruit.

   _______ c. Cycads are regarded by botanists as living fossils because they resemble extinct forms of gymnosperms.

- - - - - - - - - - - - - - - - - - -

b and c are correct

5. In Chapter 4 you studied the process of seed formation. Now, before we consider the details of reproduction in a cycad, let us summarize some of the characteristics we expect a cycad to possess as a seed plant. Identify the true statements from among the following.

_______ a. The tree-sized, seed-producing cycad is the sporophyte generation.

_______ b. A cycad may be expected to be heterosporous.

_______ c. A cycad may be expected to produce spores in sporangia which are located on sporophylls.

_______ d. It would not be surprising to find that cycads produce strobili.

\- - - - - - - - - - - - - - - - - -

All four statements are true.

6. Because the several species of cycads exhibit individual differences, we must describe them in general terms.

A common characteristic of all cycads is the occurrence of megasporophylls and microsporophylls on separate trees. Thus, one tree may have a single strobilus composed solely of microsporophylls (staminate cone) while another tree nearby will develop a strobilus made up of megasporophylls only (ovulate cone).

If you were to encounter a cycad which bore a single strobilus, you should conclude that:

_______ a. the strobilus may be composed of all microsporophylls.

_______ b. the strobilus may be composed of all megasporophylls.

_______ c. like Selaginella, the strobilus may be composed of both microsporophylls and megasporophylls.

\- - - - - - - - - - - - - - - - - -

a and b are correct conclusions

7. Each microsporophyll of a cycad has many microsporangia scattered on its lower surface. These young microsporangia contain microspores which have been produced by the process of meiosis. Does this cycad produce a staminate strobilus or an ovulate strobilus?

____

_____________ Explain your answer. _____________

____

- - - - - - - - - - - - - - - - - -

staminate, because it has a strobilus composed solely of microsporophylls

8. As in all seed plants, the microspores of the cycad are retained in the numerous microsporangia until they form microgametophytes. As the microsporangial walls burst open in a mature staminate cycad cone, dustlike particles are discharged into the air.

   a. What name is given to these dustlike particles? _____________

   b. What alternate name may be applied to these particles?

   ____

- - - - - - - - - - - - - - - - - -

a. microgametophytes (or pollen); b. pollen (or microgametophytes)

9. Using the information from frame 8 supply the correct labels for the following diagrams.

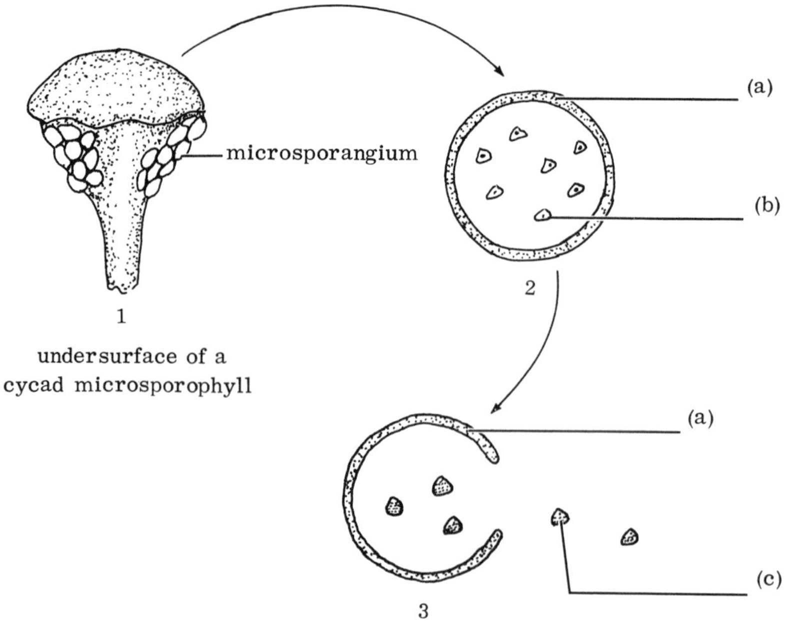

microsporangium

1

undersurface of a cycad microsporophyll

(a)

(b)

2

(a)

(c)

3

- - - - - - - - - - - - - - - - - -

a. microsporangium; b. microspore; c. microgametophyte or pollen grain

10. While the pollen grains of our hypothetical cycad are floating in the air, let us turn our attention to the ovulate strobilus which is composed solely of ____________________.

- - - - - - - - - - - - - - - - - -

megasporophylls

11. You already know that megasporophylls not only produce megasporangia, but that in a seed plant such as a cycad, the megasporangia are enclosed by enveloping integuments to form structures called (ovules/megagametophytes) ____________________.

- - - - - - - - - - - - - - - - - -

ovules

12. A megasporophyll of a cycad is a broad and stalked structure which produces two ovules on its undersurface as illustrated by these diagrams.

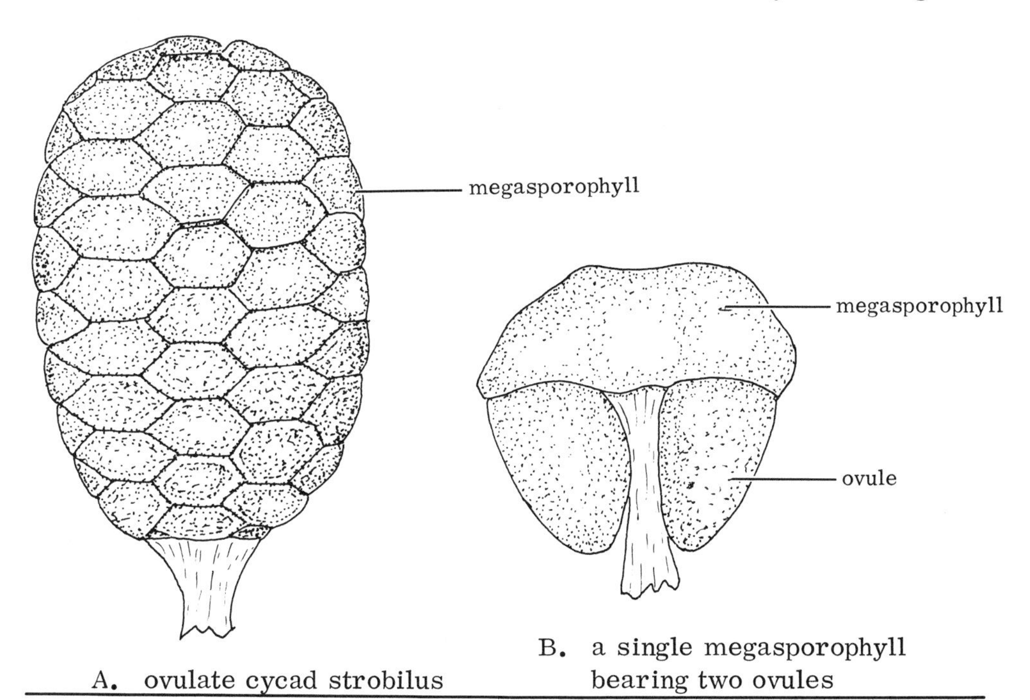

A. ovulate cycad strobilus

B. a single megasporophyll bearing two ovules

If an ovulate strobilus is composed of 50 megasporophylls, what will be the maximum number of ovules produced by that strobilus?

____

\- - - - - - - - - - - - - - - - - - -

100 ovules

13. A cycad ovule which is attached to the surface of the megasporophyll is like all ovules in that it is composed of two portions: the outer enveloping layers called the ____________________, and the central portion called the ____________________.

\- - - - - - - - - - - - - - - - - - -

integuments; nucellus (or megasporangium)

14. In Chapter 4 you learned that the megasporangium or nucellus of a seed plant produces a single functional megaspore. A single 2n megaspore mother cell divides by meiosis within the cycad nucellus to produce four megaspores which possess the haploid (n) chromosome number. Three of the four cycad megaspores disintegrate within the nucellus, leaving a single megaspore. This megaspore is called the functional megaspore.

    Check the true statement(s) regarding megaspore development in the cycads.

    _______ a. Four megaspore mother cells form megaspores within a single nucellus.

    _______ b. Each megasporangium produces one functional megaspore.

    _______ c. Four megaspores are formed within a single megasporangium.

    _______ d. Three megaspores are nonfunctional in a single nucellus.

\- - - - - - - - - - - - - - - - - - -

b, c, and d are true

15. Eventually, the single functional megaspore which is lying embedded within the cycad megasporangial tissue greatly enlarges while its nucleus undergoes a series of mitotic divisions. Thus, a multicellular megagametophyte is produced which occupies the central portion of the nucellus.

    In the series of drawings which follow, label megaspores, megaspore mother cell, megagametophyte, and functional megaspore.

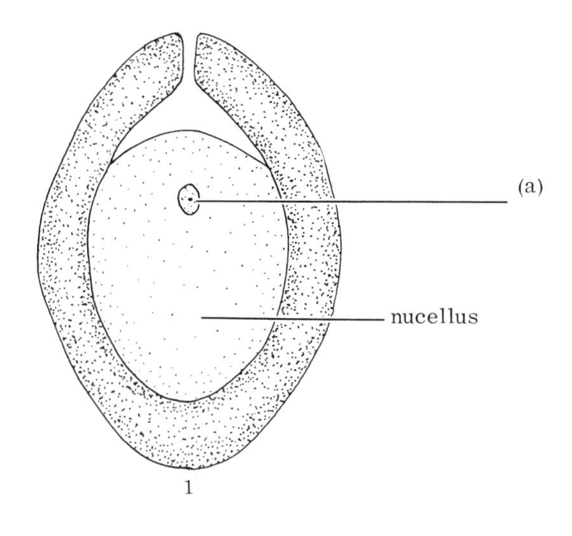

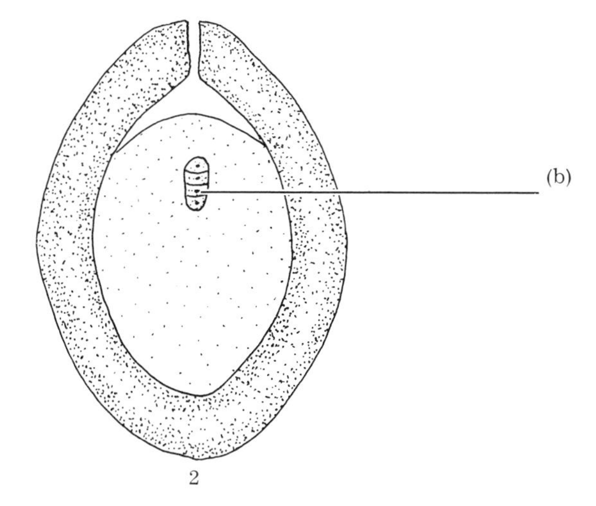

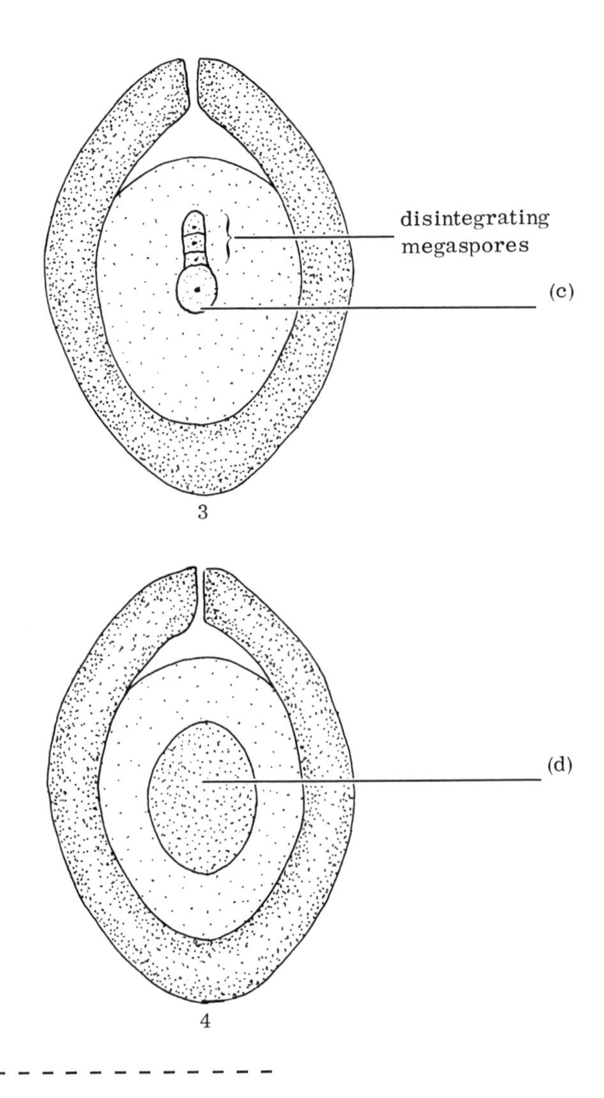

- - - - - - - - - - - - - - - - - - -

a. megaspore mother cell; b. megaspore; c. functional megaspore; d. megagametophyte

16. The megagametophyte of cycad forms several (sperms/eggs) ____________________ in the region closest to the micropyle (see diagram).

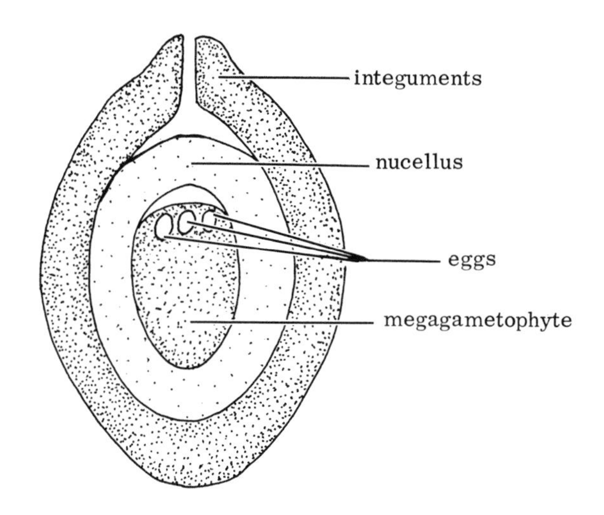

- - - - - - - - - - - - - - - - - -

eggs

17. Refer again to the diagram in frame 16. If you know that the haploid chromosome number in this species of cycad is 22, how many chromosomes will be present in the following cells of this species of cycad?

    a. integument cells _______

    b. nucellus cells _______

    c. egg cells _______

    d. megagametophyte cells _______

- - - - - - - - - - - - - - - - - -

a and b are both diploid with $2\underline{n} = 44$; c and d are haploid with $\underline{n} = 22$

18. Upon maturity of the ovule, a few cells of the nucellus in the area of the micropyle disintegrate to form a sticky fluid which exudes from the micropyle. When pollen grains are blown by the wind to the surface of the ovule, many of them adhere to the sticky fluid on the micropyle. Eventual evaporation of the fluid pulls the pollen grains down the micropyle until they come to rest on the surface of the nucellus (see diagram).

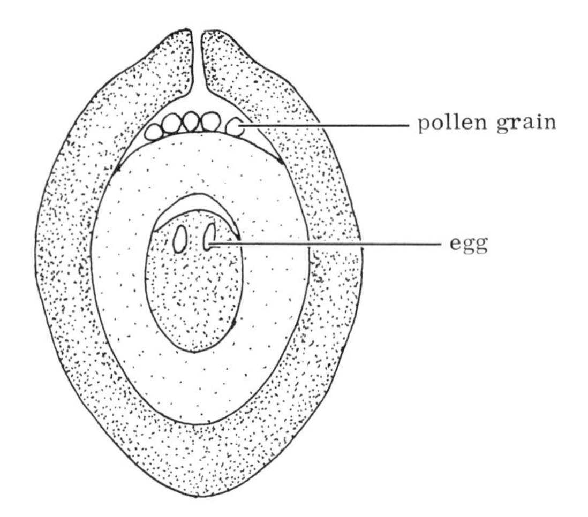

What tissue separates the microgametophytes from the megagametophyte? ____________________

\- - - - - - - - - - - - - - - - - -

nucellus (megasporangium)

19. You recall that pollen grains of seed plants penetrate the nucellus barrier which separates the two gametophytes by forming tubular outgrowths called pollen tubes. Accordingly, cycad pollen grains send out massive pollen tubes after coming to rest on the surface of the nucellus or megasporangium. As the pollen tubes of cycad grow through the nucellus, two large swimming sperms form within each tube. The presence of swimming sperms in a seed plant is considered a primitive condition characteristic of aquatic plants, and this is one of several reasons why the cycads are considered to be primitive plants (as we shall see).

    We can expect the sperms in the pollen tubes of cycads, like all seed plants, to be released

    _______ a. when the pollen grains contact the micropylar fluid.

    _______ b. as the pollen tubes grow through the nucellus.

    _______ c. when the pollen tubes completely penetrate the nucellus.

\- - - - - - - - - - - - - - - - - -

c is the correct statement

20. Because there are several eggs within a cycad megagametophyte, several embryos are formed following fertilization. Competition among the developing embryos usually results in the elimination of all but the most vigorous of the competing embryos.

The mature embryo lies in the (ovule/seed) ________________ and its food supply comes from the ________________ which completely surrounds it.

- - - - - - - - - - - - - - - - - -

seed (remember, the seed is the ovule containing an embryo); megagametophyte

21. The seed grows much larger as the embryo grows and the megagametophyte stores food produced by the parent plant. The megasporophylls comprising the ovulate strobilus are forced apart by the enlarging tissues so that the attached mature seeds lie naked and exposed to view.

The microsporophylls comprising the staminate strobilus wither and disintegrate upon completion of microspore production.

Aside from some protection afforded by the megasporophylls of an ovulate strobilus, what tissues, if any, cover and protect the cycad seeds? (Before answering, recall that the seed coats are an integral part of the seed.) ________________________________

- - - - - - - - - - - - - - - - - -

There are no coverings, hence the name, gymnosperms.

22. To summarize the life cycle of a cycad, study the following diagram and fill in the names of the missing structures.

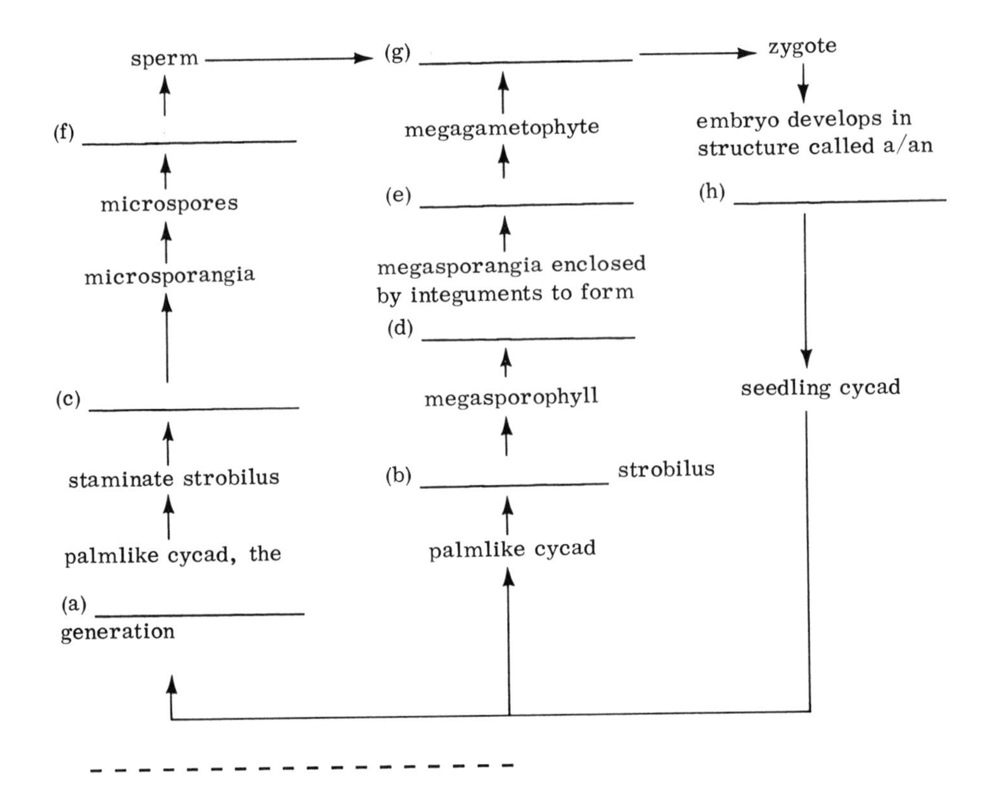

- - - - - - - - - - - - - - - - - - - -

a. sporophyte; b. ovulate; c. microsporophylls; d. ovules; e. functional megaspore; f. microgametophyte (pollen grain); g. egg; h. seed

23. Suppose that in a certain cycad the chromosome count in a leaf cell is 72. Using the diagram in frame 22 as assistance, what chromosome counts do you expect in each of the following cells of this species of cycad?

    a. cell in a microsporophyll ________

    b. megaspore ________

    c. sperm ________

    d. cell of a megasporangium ________

    e. cell in a pollen grain ________

    f. megaspore mother cell ________

    g. cell in the embryo ________

    h. functional megaspore ________

i. microspore _______

j. cell in a megasporophyll _______

- - - - - - - - - - - - - - - - - -

a. 72; b. 36; c. 36; d. 72; e. 36; f. 72; g. 72; h. 36; i. 36; j. 72

24. The cycad is a primitive gymnosperm (that is, it possesses characteristics similar to those of ferns and club mosses), but with a single exception, the remaining gymnosperm groups in existence today are more complex and more highly evolved than the cycads. Such a group is the conifers (cone-bearers), including the familiar hemlock, spruce, fir, cedar, and pine as well as several others.

    Fossil evidence suggests that the conifer line of gymnosperms is considerably older than that of the cycads, yet the cycads possess more primitive characteristics!

    On the basis of the information in this frame, which of the following statements are reasonable assumptions concerning the evolution of gymnosperms?

    _______ a. Older lines of gymnosperms are more likely to resemble ancient gymnosperms than the younger lines.

    _______ b. The cycads have evolved to a lesser degree than the pines.

    _______ c. Because the juniper is a conifer, it may be regarded as more complex and highly evolved than the cycads.

- - - - - - - - - - - - - - - - - -

b and c are reasonable assumptions concerning gymnosperm evolution

25. Although the conifers and the cycads possess similar life cycles, conifers differ in certain aspects of their life cycles. Because the pine is a familiar conifer, let's study its life cycle. Unlike the cycads, the pine tree develops both staminate and ovulate strobili on the same tree.

    Check the following statement(s) which accurately describe(s) a pine tree.

    _______ a. Every mature pine tree produces seeds.

    _______ b. Microsporophylls and megasporophylls are present in the same pine cone.

_______ c. Every mature pine tree produces pollen.

- - - - - - - - - - - - - - - - - - -

Statements a and c accurately describe the pine tree. Statement b is incorrect because the pine tree possesses two kinds of strobili—staminate cones composed of microsporophylls and ovulate cones composed of ovule-bearing structures.

26. The development of pine staminate cones and the subsequent microgametophyte development are essentially similar to that of the cycads except that the staminate cones are small and are produced in great abundance on each pine tree. The microsporangia of pine develop on the lower surface of microsporophylls and microgametophyte development is completed within the microsporangia.

    To summarize the similarities between the cycad and the pine, with respect to the organization of the staminate strobilus, place the following terms in correct sequential order from the largest and most inclusive to the smallest structures.

    (A) microspores, (B) staminate cone, (C) microsporangia, (D) microgametophytes, (E) microsporophylls.

    ____

- - - - - - - - - - - - - - - - - - -

(B) staminate cone, (E) microsporophylls, (C) microsporangia, (A) microspores, (D) microgametophytes

27. Each ovulate strobilus of pine begins its development with new spring growth at the tips of the branches of the tree. About one-fourth inch long at the time of pollination, the tiny ovulate cone differs in structure from the ovulate cycad cone. You recall that the ovulate cycad cone produces ovules on the surface of megasporophylls which are, in turn, clustered to form structures called strobili or cones.

    An ovulate pine cone presents a botanical mystery. For reasons too complex to be discussed in this textbook, the scalelike structures on which the pine ovules are borne are <u>not</u> considered to be megasporophylls, and they are given the descriptive name, <u>ovuliferous scales</u>. Moreover, a scalelike outgrowth called a <u>cover scale</u> arises below each ovuliferous scale. Because of the double nature of these strobilus elements, the ovulate cone is often described as compound. The diagrams below illustrate the relationship between the ovule, the ovuliferous scale, and the cover scale in an ovulate pine cone.

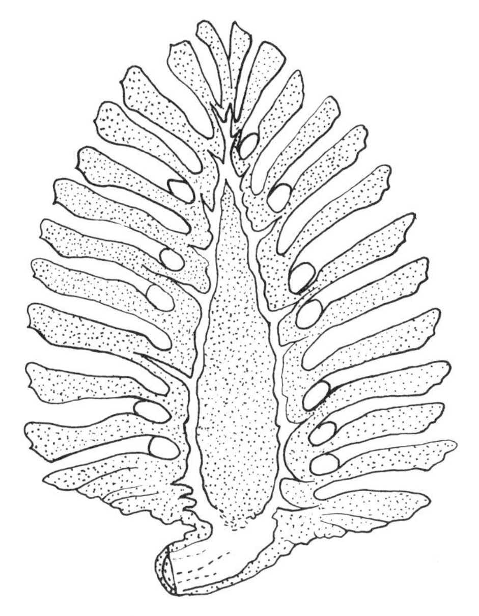

A. Open cone of gymnosperm

(from T. Elliot Weier, C. Ralph Stocking, and Michael G. Barbour, Botany: An Introduction to Plant Biology, 5th ed. (New York: John Wiley & Sons, 1974), p. 572)

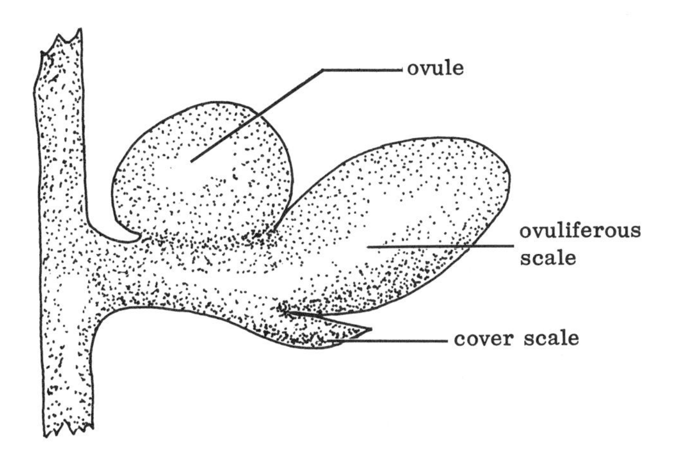

B. Naked seed within cone

Compare and contrast the composition of the cycad and pine ovulate strobili. Note why the pine ovulate strobilus is considered compound.

- - - - - - - - - - - - - - - - - -

Your answer should have included the following points: (1) Both the cycad and the pine ovulate strobilus produce ovules. (2) The ovules of the cycad are borne on megasporophylls which constitute the entire ovulate strobilus. (3) The ovules of the pine are produced on ovuliferous scales below which are cover scales, thus making the ovulate strobilus of the pine a compound structure.

28. Pollination in pine occurs in the same manner as in the cycads. The microgametophytes or pollen grains become trapped in sticky fluid exuding from the micropyle of the pine ovule. The pollen grains finally come to rest upon the surface of the nucellus or megasporangium.

Because the pollen tubes of pine grow slowly, a time lapse of nearly a year separates the two processes of pollination and fertilization.

What is happening to the pollen during the months following pollination? ______________________________

____

- - - - - - - - - - - - - - - - - -

The pollen tubes are growing through the nucellus toward the megagametophyte.

29. The two sperms formed within each pollen tube of pine are represented solely by nuclei. In other words, pine lacks mobile sperm cells which must swim through a liquid medium to reach the eggs.

The microgametophytes and the megagametophytes of seed plants do not develop on the ground near water as do those of fern and club moss which possess fewer adaptations to terrestrial life. Swimming sperms are of no advantage to a seed plant, and may be considered a primitive characteristic which has been omitted in the conifers, which have become more adapted to life on land through natural selection.

It may be reasonably concluded that extinct gymnosperms of 300 million years ago had sperms similar to those of the (pines/cycads) ____________________, because the possession of swimming sperms is considered to be a/an (advanced/primitive) ____________________ characteristic.

- - - - - - - - - - - - - - - - - - -

cycads; primitive

30. In this book we have been emphasizing the evolutionary progression of plants from primitive to more advanced forms. From the following pairs of items select the more advanced member of each pair and write its number in the blank.

_______ a. (1) gametophytes retained in sporangia; (2) free-living gametophytes on damp soil

_______ b. (1) cycad; (2) pine

_______ c. (1) non-mobile sperms; (2) swimming sperms

_______ d. (1) moss; (2) cycad

_______ e. (1) microgametophyte with sperm-conveying pollen tube; (2) microgametophyte lacking pollen tube

- - - - - - - - - - - - - - - - - - -

a. 1; b. 2; c. 1; d. 2; e. 1

31. A mature ovulate pine cone opens after two years to discharge its (naked/covered) ____________________ seeds. The familiar woody scales which comprise an ovulate pine cone are the cover scales which undergo remarkable enlargement from their inception.

If the woody structures of an ovulate pine cone are cover scales, then the ovuliferous scales should be found immediately (above/below) ____________________ the woody cover scales.

- - - - - - - - - - - - - - - - - -

naked; above

32. The pine seeds which fall from an ovulate strobilus possess elongated winglike scales which enable the wind to carry them a considerable distance from the parent tree. If the wings of the seeds are not formed by elongation of the seed coats, what other structures of the pine cone are intimately associated with the seeds and may be responsible for formation of the "wings"? ______________________________

- - - - - - - - - - - - - - - - - -

If you said "ovuliferous scales," you guessed well! The "wings" of pine seeds represent part of the ovuliferous scales.

33. The characteristics which have been given for the pine tree represent characteristics common to most conifers.

    Since a fir tree is a conifer, check the following characteristics which you would expect a fir tree to possess.

    _______ a. well-defined megasporophylls

    _______ b. well-defined microsporophylls

    _______ c. swimming sperms

    _______ d. cover scales

    _______ e. winged seeds

    _______ f. staminate and ovulate cones on separate trees

- - - - - - - - - - - - - - - - - -

b, d, and e. Note that a is not correct since the ovulate strobilus of conifers is composed of ovuliferous and cover scales.

SELF-TEST
Kinds of Seed Plants: Gymnosperms

Before you go on to the next chapter of this book, evaluate your understanding of gymnosperms by taking this short self-test. All questions should be answered in the light of information presented in Chapter 5 and preceding chapters of this book. Answers are given at the end of the test.

1. To summarize the life cycle of a pine, study the diagram below and fill in the names of the missing structures.

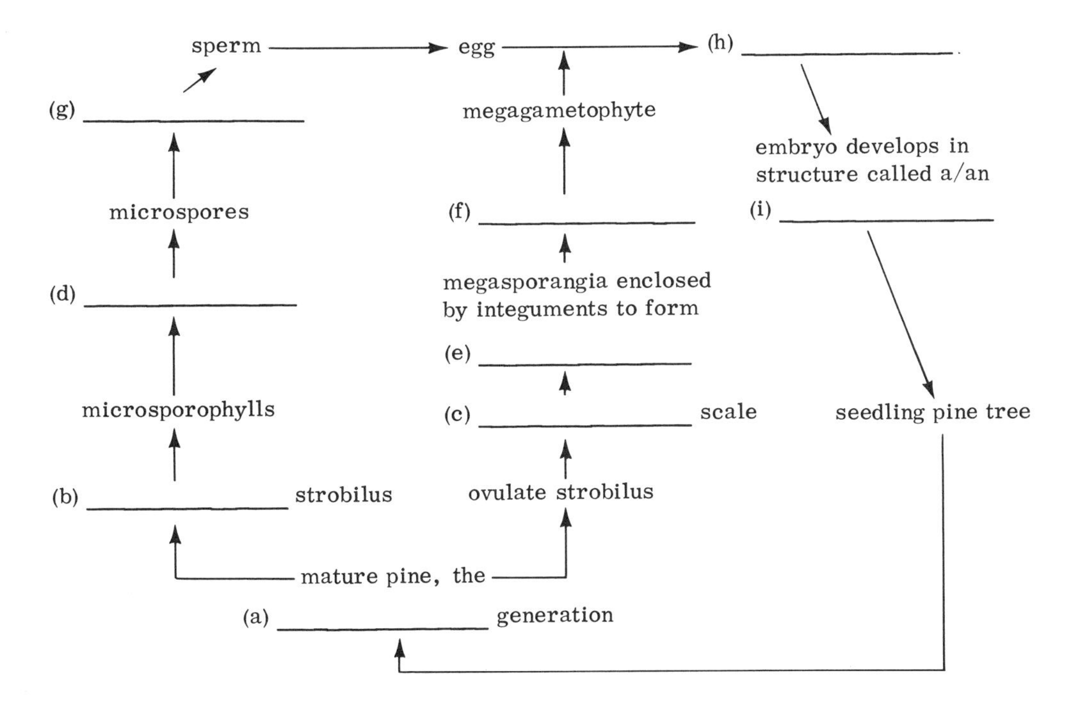

2. List the letters in the diagram in question 1 that correspond to the haploid structures. ____________________

3. On the left hand side of the diagram in question 1 (the side beginning with "(b) _____ strobilus" and ending with "sperm"), the word "meiosis" could be inserted between ____________ and ____________ to indicate the time of meiosis in that part of the pine life cycle.

4. The cycad life cycle obviously fits into the classical alternation of generations we have observed in liverworts, mosses, ferns, and Selaginella. Compare alternation of generations in these various plants by identifying and checking the true statements among the following.

   _______ a. The gametophyte is the conspicuous generation in both moss and cycad.

   _______ b. The cycad, like Selaginella, is heterosporous.

   _______ c. Both cycad and Selaginella produce strobili.

   _______ d. Both cycad and Selaginella are primitive seed plants.

   _______ e. In all of the plants—moss, liverwort, Selaginella, and cycad—the process of meiosis results in the production of spores.

5. Study the following diagrams of parts of a cycad plant. Knowing what you do about the life cycle of cycads, answer the questions about these diagrams.

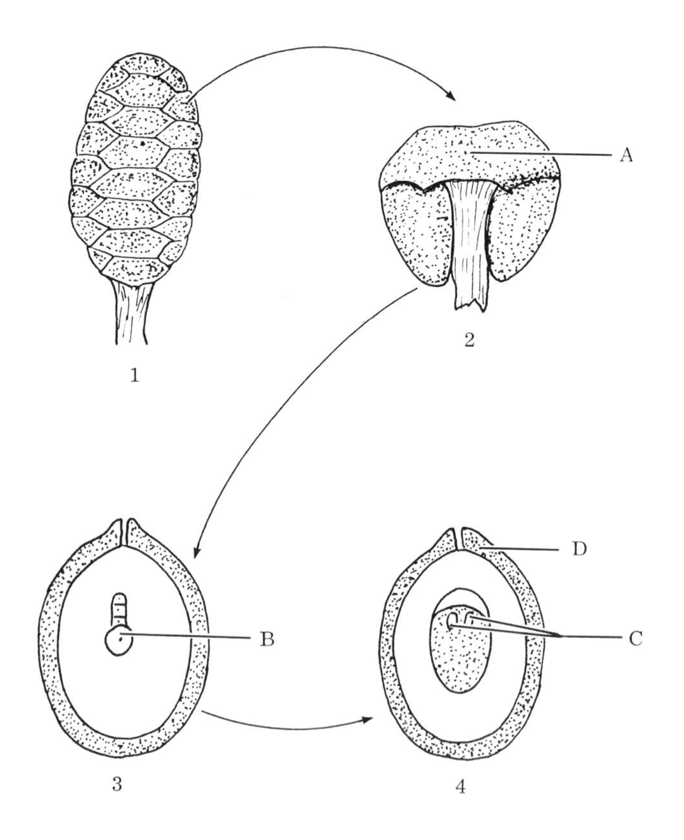

a. If the nucleus of cell B in diagram 3 contains 10 chromosomes, how many chromosomes are present in the nuclei of the cells in A of diagram 2? ____________________

b. If the nucleus of cell B in diagram 3 contains 10 chromosomes, how many chromosomes are present in the nucleus of cell C in diagram 4? ____________________

c. Diagram 1 is called what? ____________________

d. Structure A is called what? ____________________

e. Structure B is called what? ____________________

f. Structure C is one cell in a multicellular structure called what?

____

g. The entire structure numbered 4 is called a/an ______________ and it will eventually become a/an ____________________.

h. The structure labeled D is called the ____________________ and will eventually mature into the ____________________.

6. Spruce trees, which are often used as Christmas trees, are conifers, gymnosperms similar to pines and firs in their life cycles. Suppose that you determined that the nucleus of a cell in the root of a spruce tree contained 36 chromosomes. Knowing what you do about the life cycle of a conifer, answer the following questions.
How many chromosomes should be present in the nucleus

a. in a cell in a microsporophyll of this spruce? ________

b. in a cell of the megagametophyte of this spruce? ________

c. in a cell in a pollen grain of this spruce? ________

d. of a megaspore mother cell of this spruce? ________

e. of a functional megaspore of this spruce? ________

f. in a cell of an embryo in the seed of this spruce? ________

g. of a cell in one of the needlelike leaves of this spruce? ________

7. Throughout the book emphasis has been given to the evolutionary progression of plants from relatively primitive to more advanced forms. From the following pairs of items select the more primitive member of each pair and write its number in the blank.

_______ a. (1) Plants with relatively large, conspicuous gametophyte; (2) plants with large, conspicuous sporophyte.

_______ b. (1) Plant with heterospory; (2) plant with homospory.

_______ c. (1) Plant with swimming sperms; (2) plant with sperms lacking structures for locomotion.

_______ d. (1) Plants with seeds; (2) plants lacking seeds.

_______ e. (1) Plants retaining megaspores in the megasporangium; (2) plants releasing the megaspores from the megasporangium.

_______ f. (1) Cycad; (2) spruce.

8. Indicate what each of the following items becomes or develops into.

   a. megaspore mother cell _______________

   b. ovule _______________

   c. microspore _______________

   d. embryo _______________

   e. integuments _______________

   f. functional megaspore _______________

   g. zygote _______________

   h. cover scales in pine cone _______________

   i. ovuliferous scales in pine cone _______________

Answers

1. a. sporophyte; b. staminate; c. ovuliferous; d. microsporangia; e. ovules; f. megaspore; g. microgametophyte; h. zygote; i. seed (frames 25-32)

2. f and g (frames 7, 8, 14)

3. (d) microsporangia and microspores (frame 7)

4. a. false (Chapter 2, frame 3; Chapter 5, frame 5)
   b. true (Chapter 2, frame 16; Chapter 5, frame 5)
   c. true (Chapter 2, frame 16; Chapter 5, frame 5)
   d. false (Chapter 4, frame 12; Chapter 5, frame 4)
   e. true (Chapters 2 and 3; Chapter 5, frames 7, 14)

5. a. 20 (frame 23)
   b. 10 (frames 22, 23)
   c. ovulate strobilus or cone (frame 12)
   d. megasporophyll (frame 12)
   e. functional megaspore (frame 15)
   f. megagametophyte (frame 16)
   g. ovule, seed (Chapter 4, frame 35; Chapter 5, frames 12, 20)
   h. integuments, seed coats (Chapter 4, frame 37; Chapter 5, frames 11, 16)

6. a. 36; b. 18; c. 18; d. 36; e. 18; f. 36; g. 36 (frames 22, 23)

7. a. 1 (frames 29, 30)
   b. 2 (Chapter 4, frame 11)
   c. 1 (frames 29, 30)
   d. 2 (frame 1)
   e. 2 (Chapter 4, frame 12)
   f. 1 (frame 24)

8. a. four megaspores (frame 14)
   b. seed (frame 20)
   c. microgametophyte (pollen grain) (frame 8)
   d. seedling sporophyte and eventually mature sporophyte (frame 22)
   e. seed coats (Chapter 4, frame 37)
   f. megagametophyte (frame 15)
   g. embryo (frame 22)
   h. woody scales of mature ovulate pine cone (frame 31)
   i. winglike projections on pine seeds (frame 32)

CHAPTER SIX

Kinds of Seed Plants: Angiosperms

1. As we have already seen, the earliest seed plants were gymnosperms, and they remained the only seed plants for millions of years until the flowering plants came into existence to compete with them. The presence of flowers is an outstanding characteristic of this group of seed plants, and their seeds are completely enclosed by a portion of the flower so that the term angiosperms ("covered seeds") has been applied to them.

   Two outstanding characteristics possessed by angiosperms are

   ____________________ and ____________________.

   \- - - - - - - - - - - - - - - - - -

   flowers; covered seeds (either order)

2. The evolutionary origin of the angiosperms is a mystery, one that Charles Darwin stated he would like to see solved. Despite the lack of positive evidence, almost all botanists assume that some unidentified ancestral gymnosperm evolved into the first angiosperm. Whether angiosperms evolved from gymnosperms or not, the following discussion of flowering plants is based on this assumption. Many similarities between angiosperms and gymnosperms exist, and the angiosperms may be better understood when parallel developments between the two groups are stressed.

   The morphological nature of angiosperms seems to indicate that (covered/naked) ____________________ seeds have been derived from the more primitive (covered/naked) ____________________ seed condition.

   \- - - - - - - - - - - - - - - - - -

covered; naked

3. Since the presence of flowers is a unique characteristic of angiosperms, let us begin by the discussion of flower parts. The enlarged tip of a flower stalk to which the parts of a flower are attached is termed the receptacle (see diagram). A typical flower such as that of a tomato plant has four kinds of specialized structures attached to the receptacle in layers or whorls.

   The specialized flower parts are attached in layers or ____________________ to the swollen tip of the flower stalk or ____________________.

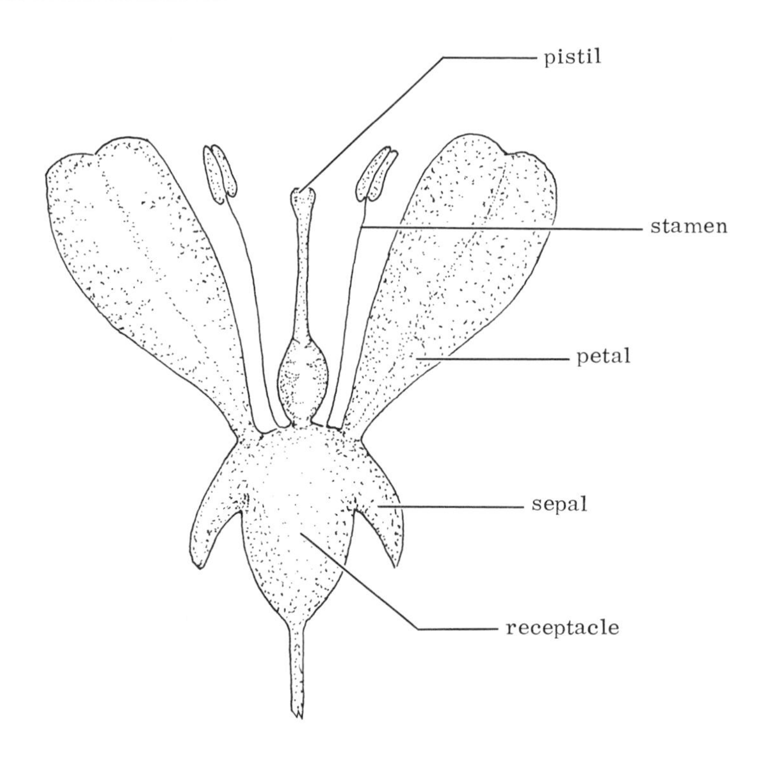

- - - - - - - - - - - - - - - - - -

whorls; receptacle

4. Flower parts in the lowermost whorl are often, but not invariably, leaflike. They are the sepals which protect the unopened flower, and they may vary widely in number and appearance (see diagram in frame 3).

Attached to the receptacle directly above the sepals is a whorl of petals which, like the sepals, may vary in appearance and number. When they are brightly colored as in the tomato flower, petals are thought to attract insects which play a role in pollination.

A spring wild flower, Rue Anemone, possesses a single whorl of white petal-like flower parts instead of the expected two layers or whorls. Careful study of the early development of the flower bud reveals that the second whorl of flower parts remains undeveloped in this plant species. Therefore, one may conclude that the flower of Rue Anemone possesses showy white ____________ which resemble the nonexistent ____________.

- - - - - - - - - - - - - - - - - -

sepals; petals

5. The third whorl of flower parts consists of the familiar stamens (see diagram in frame 3). Although varying greatly in appearance, stamens possess two main parts: a stalked portion termed the filament and the terminal knobbed end called the anther. The stamens produce pollen and have a vital function in the production of seeds. We will discuss this in detail later in this chapter.

   Label the anther and filament in the following drawings of stamens.

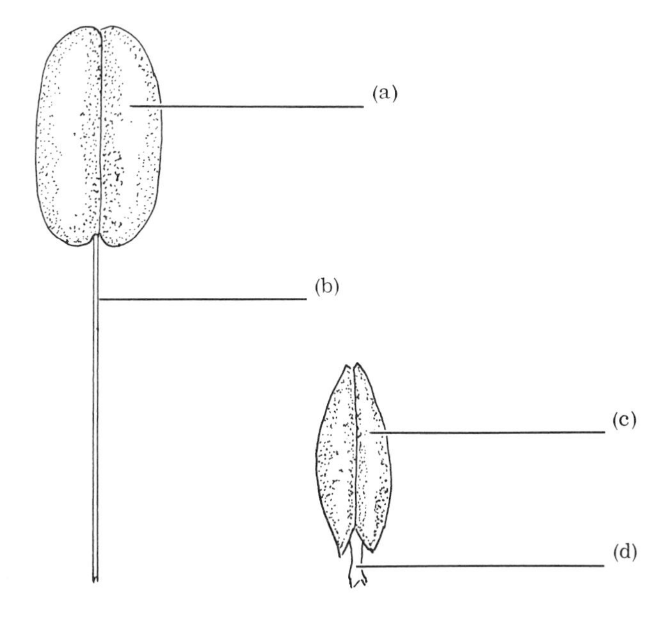

- - - - - - - - - - - - - - - - - -

a. anther; b. filament; c. anther; d. filament

6. The fourth and final whorl of flower parts is composed of pistils (see diagram in frame 3). Although pistils occur in whorls in some flowers, most flowers possess only one pistil, so that this whorl is non-existent in these flowers. Seeds will develop in the ovary portion of the pistil and this process will be described in detail later in this chapter. The pistil, so called because it was thought to resemble a pestle which is used to grind materials in a mortar, is composed of three distinct parts. Its enlarged or swollen base is called an ovary—a misnomer, the more slender necklike portion rising above the ovary is called the style, and the tip of the style is the stigma.

   Label the parts of a pistil in the following drawing.

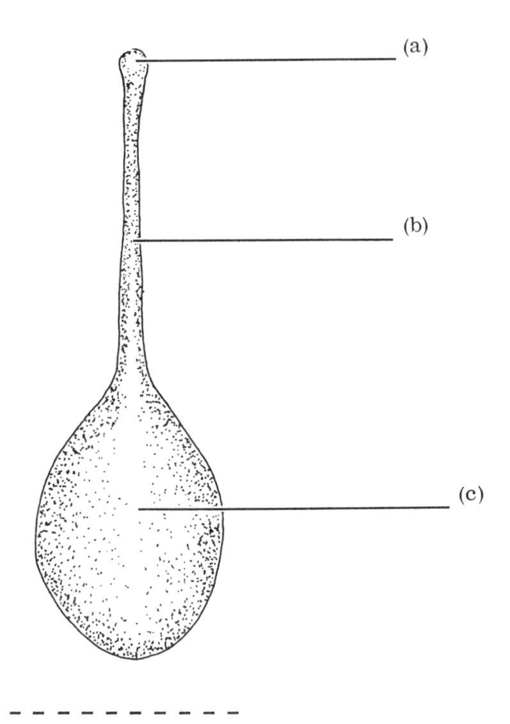

- - - - - - - - - - - - - - - - - -

a. stigma; b. style; c. ovary

7. Recall that in frame 1 of this chapter flowering plants were called angiosperms because their seeds are covered or enclosed in a portion of the flower.

   If ovules are produced in the ovary portion of the pistil, can you guess what portion of the flower encloses the seeds? ______________

- - - - - - - - - - - - - - - - - -

the ovary

8. Now, to review and summarize the parts of a flower, label the receptacle, pistil, petal, stamen, and sepal in the following drawing of a flower.

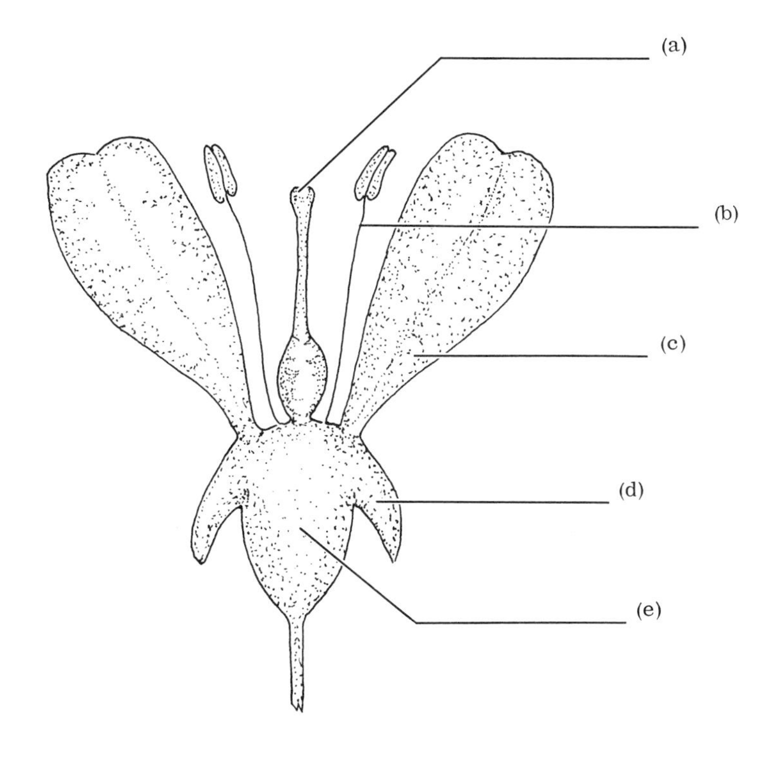

- - - - - - - - - - - - - - - - - -

a. pistil; b. stamen; c. petal; d. sepal; e. receptacle

9. The primary function of a flower is to reproduce a plant by the production of seeds. Since a flower produces seeds even when both the sepals and petals are missing, sepals and petals can be called the nonessential parts of a flower.

   Since the petals of Rue Anemone are nonexistent, and all of the other possible flower parts are present, can the flowers of Rue

Anemone be expected to produce seeds? ____________________

- - - - - - - - - - - - - - - - - -

yes

10. On the other hand, both stamens and pistils are necessary for seed production, so they are called the essential parts of a flower.

    Which of the plant mutations listed below would be most detrimental to the survival of the plant having the mutation?

    _______ a. A type of mutation in which a plant would be unable to produce flowers with sepals and petals.

    _______ b. A type of mutation in which a plant would be unable to produce flowers with stamens and pistils.

- - - - - - - - - - - - - - - - - -

The condition described in b would prevent the plant from reproducing sexually. Although the plant could survive by asexual reproduction, the plant would be unable to adapt to changing conditions as described in Chapter 1.

11. You may well ask, "Why are the stamens and pistils essential for seed production?" The answer may be obtained by a microscopic examination of these organs.

    If you microscopically examine a cross section of the anther of a very young stamen, you will see that certain cells in four definite areas of the anther are dividing by meiosis to produce a large number of microspores. We can call the areas in which the diploid cells are dividing by meiosis (microsporophylls/microsporangia) ____________________ and the cells formed by meiosis ____________. We can consider the entire stamen a (microsporophyll/microgametophyte) ____________________ which possesses four ________________.

- - - - - - - - - - - - - - - - - -

microsporangia; microspores; microsporophyll; microsporangia

12. So it seems that the microsporophylls of cycads and conifers function

in the same manner as what part of a flower? ____________________

- - - - - - - - - - - - - - - - - - - -

the stamens

13. Further examination of older stamens reveals that the four microsporangia of a single stamen enlarge and fuse in pairs to form two large chambers or pollen sacs in which the microspores have developed into microgametophytes.

    Label microsporangium, pollen sac, microspore, and pollen in the following drawings of cross sections of anthers.

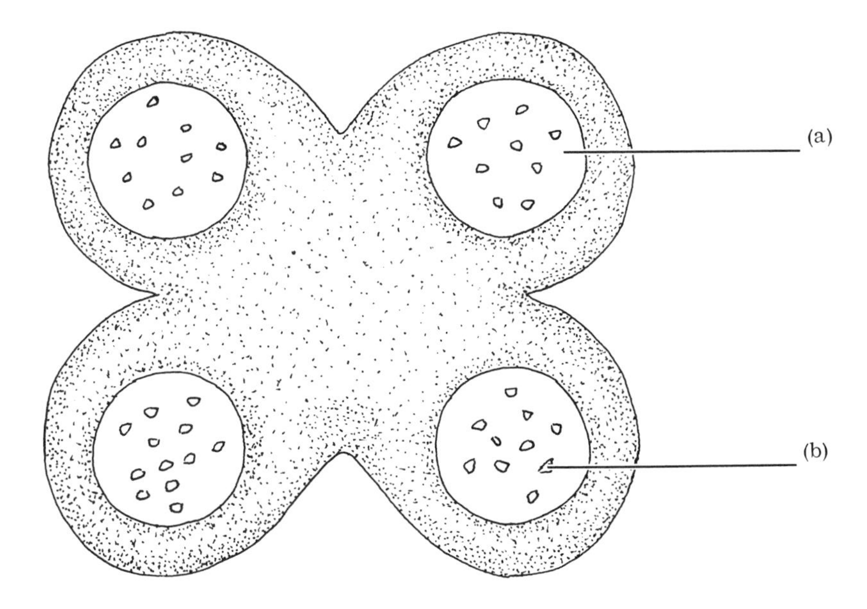

young anther

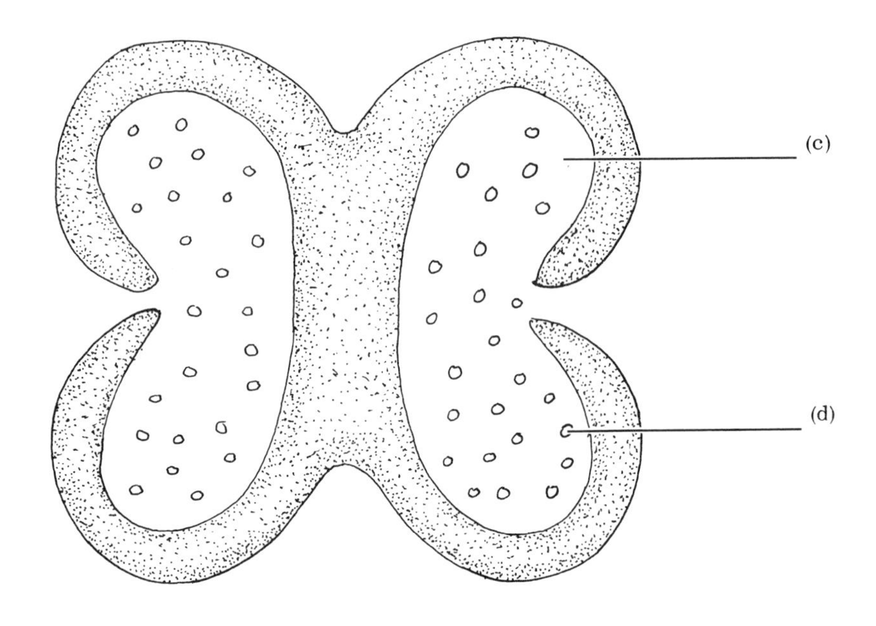

mature anther

- - - - - - - - - - - - - - - - - - -

a. microsporangium; b. microspore; c. pollen sac; d. pollen

14. Now turn your attention to the nature of the pistil. If we were to dissect a pistil, we would find that the stigma and style are made of essentially solid tissue, but the ovary is composed of one or more hollow spaces in which ovules are produced.

    If we ignore these unique parts of the pistil, and concentrate instead on the fact that the ovules (megasporangia and integuments) within the pistil are attached to the ovary wall, it should be clear that the pistil is a (megasporangium/megasporophyll) ________________.

- - - - - - - - - - - - - - - - - - -

megasporophyll. Note that ovules, modified sporangia, are always attached to sporophylls.

15. To justify the view that a pistil is really a modified megasporophyll, we must speculate on the appearance of a hypothetical gymnosperm ancestor which may have evolved a pistil-like structure. We need only

envision a gymnosperm which possessed a broad leaflike megasporophyll with ovules borne on its edges (see figure A) to understand how the pistil may have evolved.

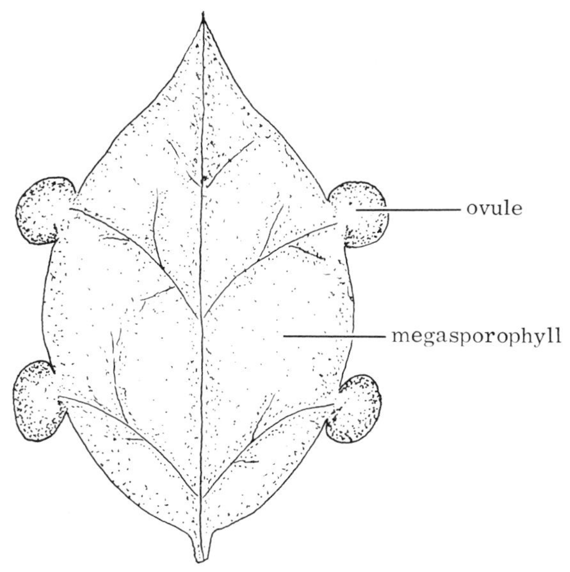

A. uncurled megasporophyll of a hypothetical gymnosperm

What we are suggesting, then, is that the broad, leaflike megasporophyll of some ancestral gymnosperm, along with the attached ovules on its edges ultimately gave rise to the pistil of flowering plants or angiosperms. Botanists believe that over a long period of time, the edges of the megasporophyll curled inwardly so that its ovules were enclosed by the enfolding megasporophyll as illustrated by figures B and C, which follow.

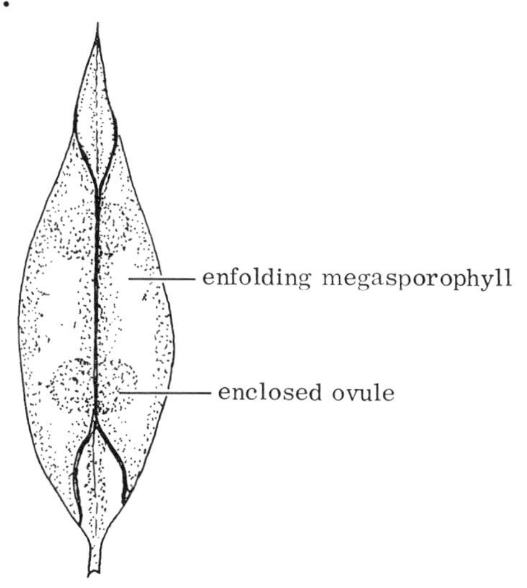

B. side view of curled megasporophyll enclosing ovules

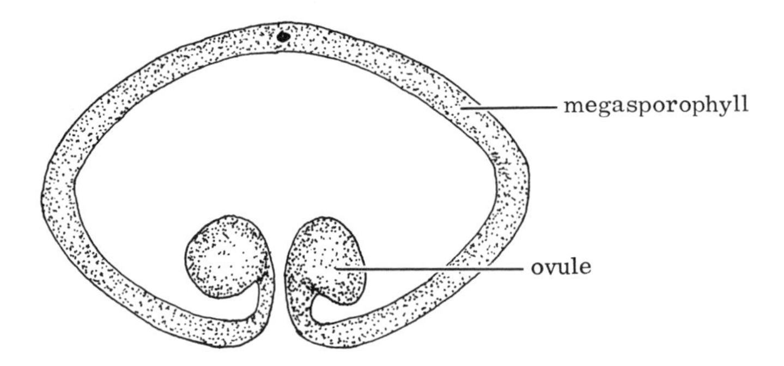

C. cross section of a curled megasporophyll shown in figure B

Because the seeds produced by such a plant are covered by the enveloping megasporophyll, this plant would now be described as a/an (angiosperm/gymnosperm) ____________________.

- - - - - - - - - - - - - - - - - - -

angiosperm

16. It is easy to visualize how such a modified megasporophyll could become a pistil with only slight further modifications, as indicated in the following drawings.

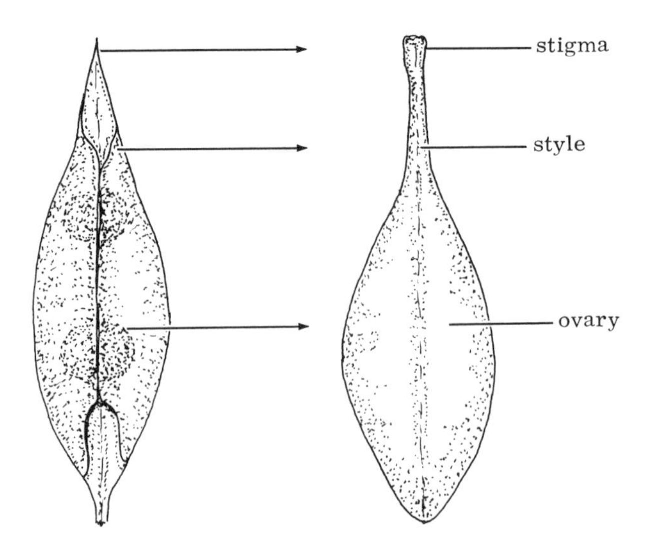

A group of plants was recently discovered in the Fiji Islands with such imperfectly closed pistils that the seeds were actually exposed to view. If you were a botanist investigating this plant group, would you consider it an advanced or primitive angiosperm group?

___________________ Why? ___________________________________

____

____

- - - - - - - - - - - - - - - - - -

You should probably regard this group as primitive. Standing intermediate between the gymnosperms, on the one hand, and typical angiosperms on the other, these plants possessing other angiosperm characters would appear to be a "missing link" in the evolutionary development of modern angiosperms.

17. Although a pistil is often composed of a single megasporophyll, frequently several megasporophylls may become intimately fused to comprise a single pistil. A pistil composed of a single megasporophyll is called a <u>simple</u> pistil, while one consisting of several megasporophylls is called a <u>compound</u> pistil.

    The Easter lily possesses three distinct stigmas on a common style. The pistil of the Easter lily is apparently composed of three megasporophylls which have completely fused except in the region of the stigmas. Is the Easter lily pistil single or compound?

    ___________________ Explain. ________________________________

    ____

    - - - - - - - - - - - - - - - - - -

    Compound, because the three stigmas reveal that the pistil is composed of three fused megasporophylls.

18. We have previously mentioned that, through common usage, inappropriate terms are often bestowed upon the various organs of seed plants. You learned that a megasporangium is commonly called a/an

    ___________________, that a microgametophyte is called a/an

    ___________________, and that a microsporophyll is sometimes given the name ______________________.

- - - - - - - - - - - - - - - - - - -

nucellus; pollen grain; stamen

19. In the same way the megasporophylls of angiosperms are usually called carpels. Therefore, the Easter lily pistil with its three stigmas may be described as 3-carpellate.

    The following drawings indicate the 3-carpellate nature of the Easter lily ovary.

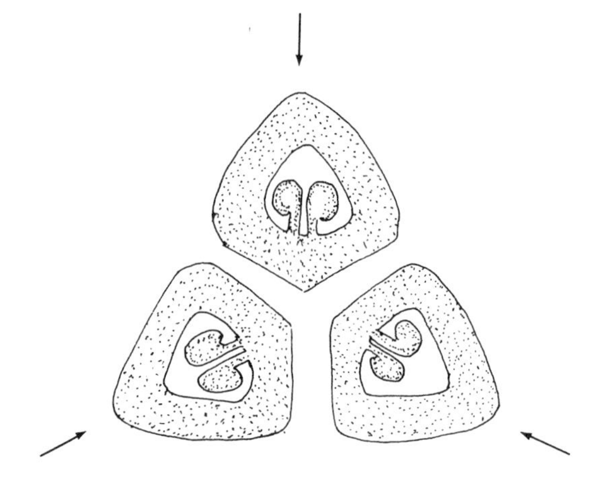

A. cross sectional view of three separate carpels

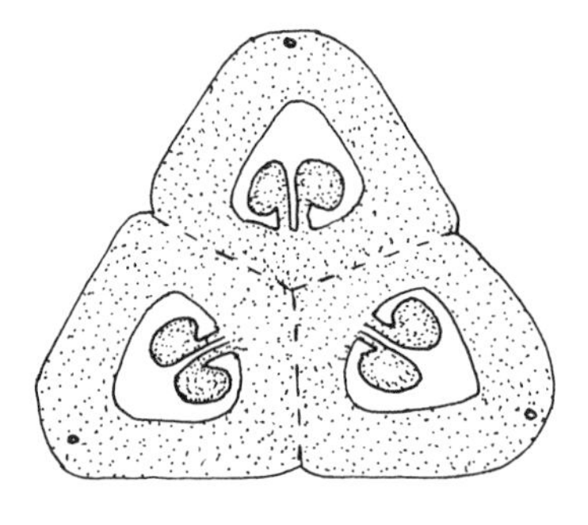

B. cross sectional view of an ovary of Easter lily showing its 3-carpellate nature

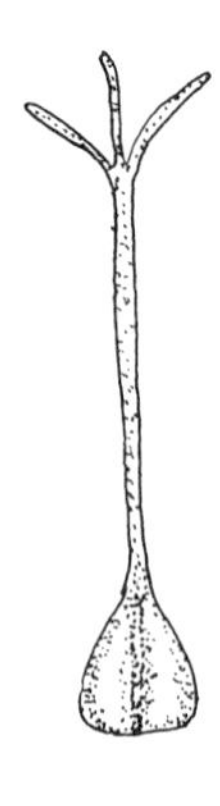

C. side view of an Easter lily pistil with fusion of the three carpels complete except in the region of the stigma

Under the following diagrams of flower pistils, indicate the number of carpels or megasporophylls involved in forming each of the pistils.

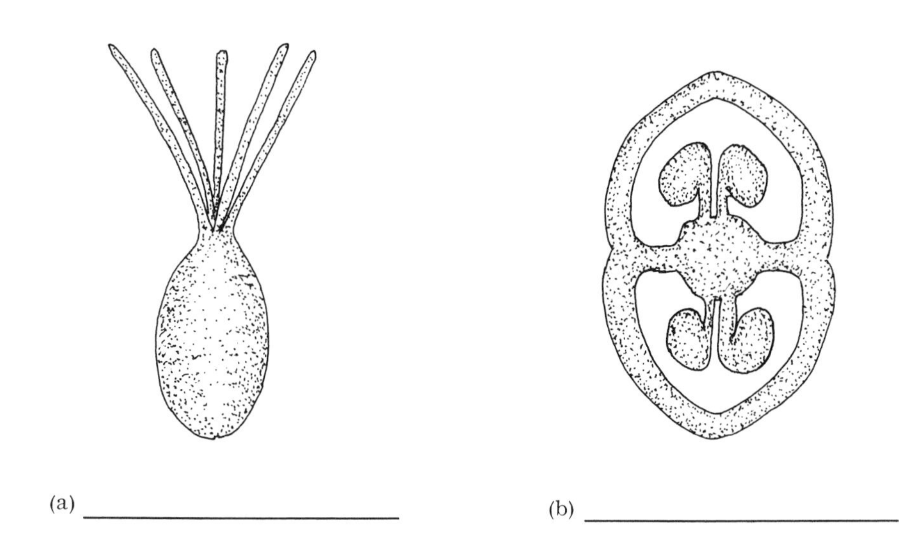

(a) ____________________ (b) ____________________

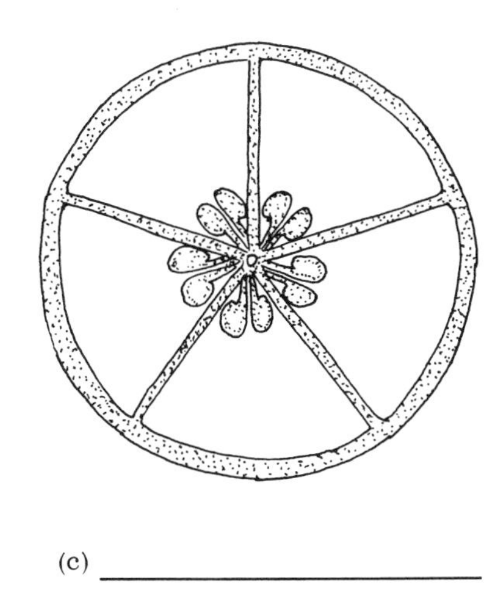

(c) ____________________

- - - - - - - - - - - - - - - - - - -

a. 5 carpels; b. 2 carpels; c. 5 carpels

20. You have learned that angiosperm stamens are comparable to the ____________________ of a gymnosperm, and that the angiosperm

pistil is composed of one or more ____________________.

\- - - - - - - - - - - - - - - - - -

microsporophylls; megasporophylls

21. If we disregard the presence of the nonessential flower parts (sepals and petals), the angiosperm flower can be regarded as a modified strobilus because a strobilus is an aggregation of ________________.

\- - - - - - - - - - - - - - - - - -

sporophylls

22. Angiosperm flowers are comparable to gymnosperm strobili because both are composed of groups of ____________________ and ____________________.

\- - - - - - - - - - - - - - - - - -

megasporophylls; microsporophylls (either order)

23. Although some flowers possess only stamens or only pistils, the angiosperm flower is usually composed of <u>both</u> microsporophylls which bear microsporangia and megasporophylls which bear megasporangia. Therefore, the flower is often described as a <u>modified bisporangiate strobilus</u>.

    A bisporangiate strobilus is one which produces two kinds of sporophylls. Some of its sporophylls bear ____________________ and some bear ____________________.

\- - - - - - - - - - - - - - - - - -

microsporangia; megasporangia (either order)

24. If the angiosperm flower has evolved from a gymnosperm strobilus, as most botanists believe, then you may ask, "Are there any gymnosperms which produce bisporangiate strobili?" None exist today. In the cycads and conifers the megasporophylls and the microsporophylls

are produced on separate strobili. However, the fossil record shows that gymnosperms with bisporangiate strobili existed in past ages.

The gymnosperm ancestor which gave rise to the flowering plants probably possessed strobili which were ____________________.

\- - - - - - - - - - - - - - - - - -

bisporangiate

25. Flowering plants are called angiosperms because their seeds are covered by the enfolded ____________________. This condition contrasts sharply with that in the gymnosperms which, as the name implies, have ____________________ seeds.

\- - - - - - - - - - - - - - - - - -

megasporophylls; naked or uncovered seeds

26. You should not conclude, however, that an angiosperm is merely a kind of gymnosperm with covered seeds, because another important difference separates the two groups. The megagametophytes and microgametophytes of angiosperms are composed of fewer cells than those of gymnosperms, and consequently are less complex. The evolutionary pattern of development of the angiosperms from gymnosperms has resulted not only in greater protection for angiosperm seeds, by covering them with modified megasporophylls or carpels, but also in the simplification of the gametophytes.

The mature ungerminated microgametophytes or pollen grains of angiosperms are composed of only <u>two cells</u> in contrast to a greater number of cells in the pollen grains of gymnosperms.

Easter lily, geranium, and petunia are well known flowering plants. In the following list, check those plants whose ungerminated pollen grains are composed of <u>more</u> than two cells.

_______ a. cycads

_______ b. Easter lily

_______ c. petunia

_______ d. fir

_______ e. pine

_______ f. geranium

- - - - - - - - - - - - - - - - - -

a, d, and e

27. The megagametophyte of an angiosperm has undergone marked reduction in complexity. In contrast to the multicellular gymnosperm megagametophyte which is composed of hundreds of cells and produces several eggs, the angiosperm megagametophyte upon maturity consists of only a few cells and contains only one egg.

    Which of the following statements accurately reflect the condition of angiosperm gametophytes?

    _______ a. Megagametophytes of angiosperms are less complex than those of gymnosperms.

    _______ b. Pollen grains of angiosperms are composed of eight cells.

    _______ c. Angiosperm megagametophytes do not contain several eggs.

    _______ d. Angiosperm microgametophytes are less complex than those of the gymnosperms.

- - - - - - - - - - - - - - - - - -

a, c, and d

28. The functional megaspore in the nucellus of a typical angiosperm undergoes a series of mitotic divisions to produce an <u>eight-nucleate</u> megagametophyte. Therefore, the mature angiosperm megagametophyte assumes the appearance of an enlarged, watery cell containing eight nuclei.

    Because of the saclike appearance of the megagametophyte and because it is the structure in which an embryo eventually develops, the angiosperm megagametophyte is called an <u>embryo sac</u>. The embryo sac may be considered a megagametophyte comparable to that of a gymnosperm because it has been formed from a functional megaspore which is lying in a megasporangium, and because its eight nuclei bear the haploid number of chromosomes.

    Compare the gymnosperm megagametophyte with the angiosperm embryo sac, with respect to their <u>origin</u> and <u>chromosome number</u>.

____

- - - - - - - - - - - - - - - - - - -

The angiosperm embryo sac is in fact a megagametophyte. Both the gymnosperm megagametophyte and the angiosperm embryo sac are formed from a functional megaspore by the process of mitosis and possess the haploid chromosome number.

29. Although the pattern of development may vary in detail, a typical embryo sac has eight nuclei arranged in a predictable fashion: an egg together with two nuclei of uncertain function assume a position in the portion of the embryo sac which is closest to the micropyle; three nuclei lie in the portion of the embryo sac farthest from the micropylar end; and two polar nuclei approach each other in the center of the embryo sac.

    Label the micropyle, nucellus, integuments, embryo sac, egg, and polar nuclei in the following drawing of an angiosperm ovule.

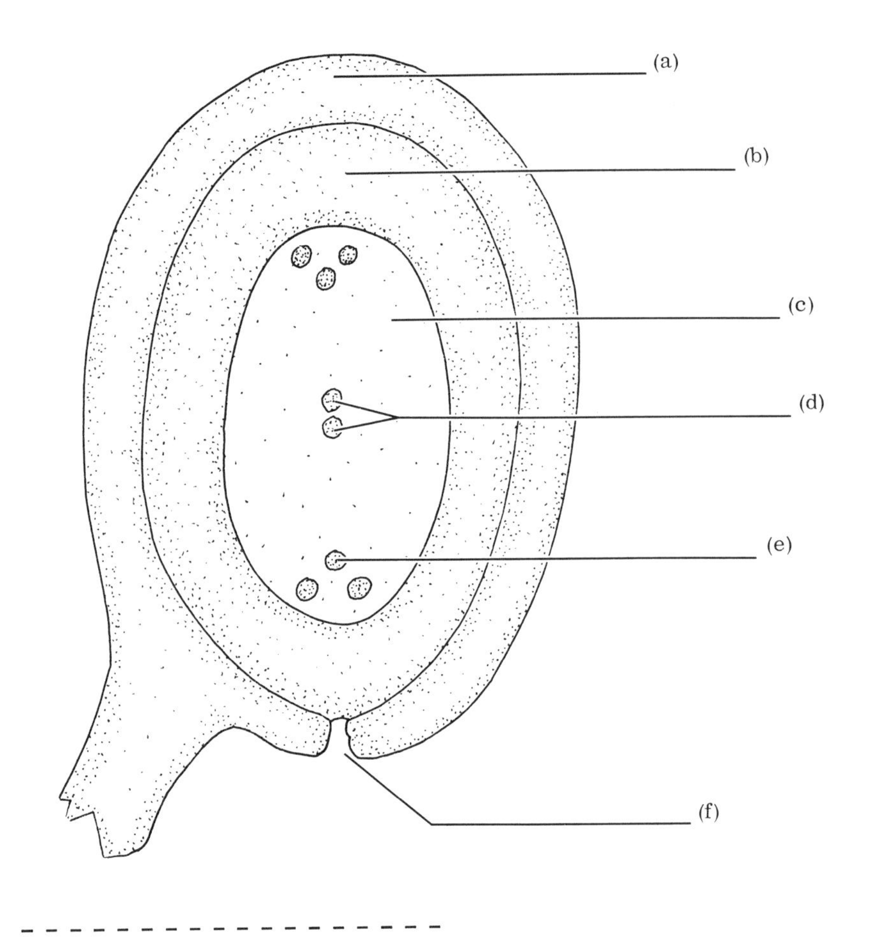

- - - - - - - - - - - - - - - - - - -

a. integuments; b. nucellus; c. embryo sac; d. polar nuclei; e. egg; f. micropyle

30. Pollination, which is easily accomplished in gymnosperms, is complicated by the presence of the pistil in angiosperms. Since angiosperm ovules are enclosed by the portion of the pistil called the

____________________, there is no opportunity for pollen grains to come in direct contact with the micropyles of the ovules.

To refresh your memory of the location of the ovules of an angiosperm, label the ovule, ovary, micropyle, stigma, style, embryo sac, and egg in the following diagram.

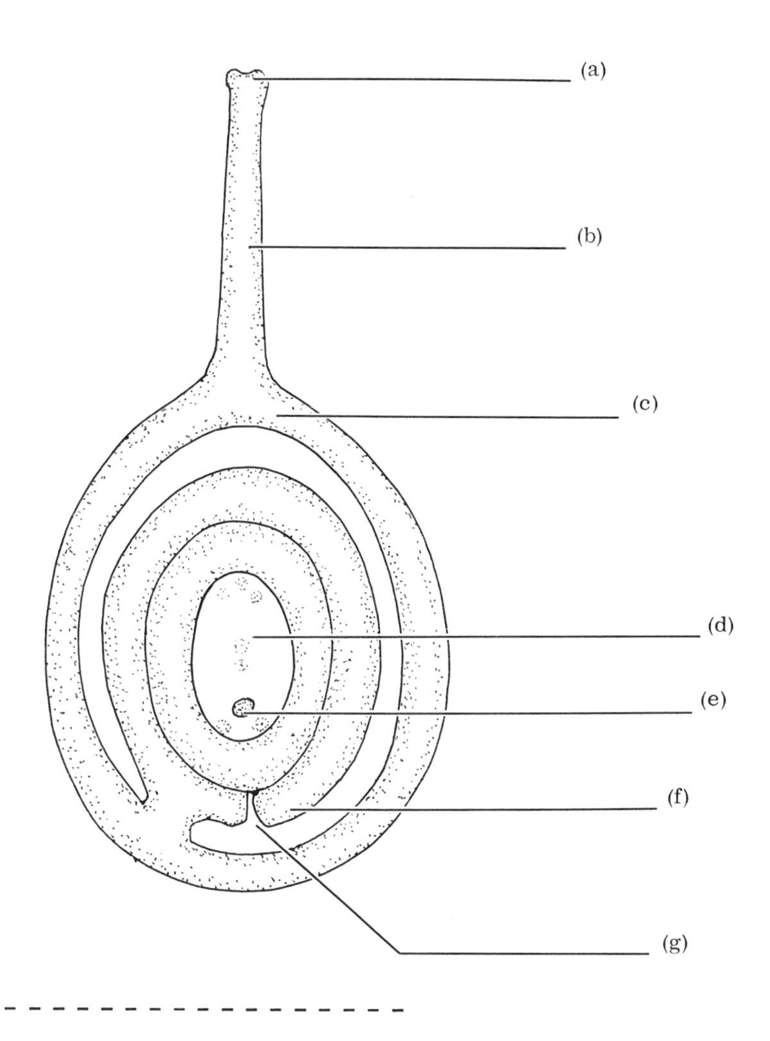

- - - - - - - - - - - - - - - - - - -

a. stigma; b. style; c. ovary; d. embryo sac; e. egg; f. ovule; g. micropyle

31. By examining the drawing of the pistil and ovule in frame 30, you can see that pollination in angiosperms results in the pollen grains coming to rest somewhere on the surface of the pistil. Only the pollen grains which lie on the stigma are capable of germinating pollen tubes.

Recall from Chapter 5 that in gymnosperms the pollen is trapped in a drop of sticky fluid exuding from the micropyle and that evaporation of that fluid results in the pollen grains coming to rest on the surface of the megasporangium (nucellus).

In angiosperms, however, pollination results in the deposition of pollen on the stigma portion of the pistil which is a (megagametophyte/

modified megasporophyll) ____________________.

- - - - - - - - - - - - - - - - - - -

modified megasporophyll

32. The tube formed by a gymnosperm pollen grain needs to be only long enough to penetrate the nucellus. The pollen tube of an angiosperm, on the other hand, must penetrate the stigma and style of the pistil and grow over the surface of the ovule before it encounters and enters the micropyle on the underside of the ovule.

    a. What passes through the micropyle of a gymnosperm ovule?

    ____

    b. What passes through the micropyle of an angiosperm ovule?

    ____

- - - - - - - - - - - - - - - - - - -

a. the entire pollen grain; b. only the pollen tube of a single pollen grain

33. In Chapter 5 you learned that the non-swimming sperm nuclei produced by the pine microgametophyte represented an evolutionary advancement in tree-sized terrestrial plants. Therefore, would you anticipate that the two male gametes produced by an angiosperm pollen grain will be non-mobile sperm nuclei? ____________________

    Explain your answer. ______________________________

    ____

- - - - - - - - - - - - - - - - - - -

Yes. Since botanists generally agree that angiosperms are well adapted to life on land, we can expect that their sperms (like those of the pine) will lack mobility.

34. The angiosperm pollen tube upon growing through the nucellus penetrates the embryo sac or megagametophyte and discharges its two sperm nuclei into the sac.

    Answer each of the following questions with an appropriate number.

_______ a. How many embryo sacs are located in each angiosperm ovule?

_______ b. How many pollen tubes penetrate the micropyle of an ovule?

_______ c. How many eggs are produced by an embryo sac?

_______ d. How many sperm nuclei are discharged into an embryo sac?

- - - - - - - - - - - - - - - - - -

a. 1; b. 1; c. 1; d. 2

35. Some time prior to the discharge of the two sperm nuclei into the embryo sac by the pollen tube, the two polar nuclei lying in the center of the embryo sac contact each other and fuse to form a fusion nucleus.
    Since each of the eight nuclei in an embryo sac contains the _________________ number of chromosomes it follows that the fusion nucleus must have the _________________ number of chromosomes.

- - - - - - - - - - - - - - - - - -

haploid; diploid

36. One of the two sperm nuclei discharged into the embryo sac contacts and fertilizes the egg nucleus to produce a zygote with the _________________ chromosome number. The second sperm nucleus fuses with the (haploid/diploid) _________________ fusion nucleus to form an endosperm nucleus with _________________ sets of chromosomes.

- - - - - - - - - - - - - - - - - -

diploid; diploid; three

37. Two nuclei in an embryo sac which are contacted by sperm nuclei are the _________________ and the _________________. If *n* represents the haploid chromosome number and 2*n* the diploid number, then

what chromosome number would you expect to find in the endosperm nucleus? ____________________

- - - - - - - - - - - - - - - - - - -

egg, fusion nucleus (either order); the 3n (or triploid) number

38. The activity of two sperm nuclei in fusing with two of the nuclei in the embryo sac is called double fertilization and the process is peculiar to angiosperm development.

    Two nuclei formed as the result of double fertilization in an embryo sac are the ________________ nucleus and the ________________ nucleus.

- - - - - - - - - - - - - - - - - - -

zygote; endosperm (either order)

39. Upon fertilization of the egg in the embryo sac, the resulting zygote is capable of developing into an embryo, provided there is sufficient food present in the ovule.

    In a gymnosperm ovule, the embryo develops within the surrounding (megasporangium/megagametophyte) ____________________, which provides an excellent source of food.

    The megagametophyte within a gymnosperm ovule serves two functions: it produces (eggs/sperms) ____________________, and it provides food for the young ____________________.

- - - - - - - - - - - - - - - - - - -

megagametophyte; eggs; embryo

40. The angiosperm megagametophyte or embryo sac consists of only eight nuclei. Such a greatly reduced megagametophyte, unlike that of a gymnosperm, possesses no stored food and is incapable of providing nourishment for its embryo.

    Check the one condition among the following which prevents the embryo sac from functioning in the manner of a gymnosperm megagametophyte.

_______ a. Embryo sacs lack several eggs.

_______ b. Embryo sacs are composed of haploid cells.

_______ c. Embryo sacs are composed of only a few cells and possess no stored food.

- - - - - - - - - - - - - - - - - -

c. Although a is true, the presence of only one egg does not prevent the embryo sac from functioning as a megagametophyte; b also is true, since all gametophytes are haploid.

41. The embryos of angiosperms survive, in spite of the inability of the embryo sac to provide adequate nourishment, due to the activity of the endosperm nucleus. Recall that the endosperm nucleus arose as the result of the fusion of two __________________ nuclei which in turn fused with a __________________ nucleus. Thus, the endosperm contains a (haploid/diploid/triploid) __________________ number of chromosomes as the result of the fusion of three (haploid/diploid/triploid) __________________ nuclei.

- - - - - - - - - - - - - - - - - -

polar; sperm; triploid; haploid

42. The endosperm nucleus undergoes repeated mitotic divisions to produce a multicellular tissue which completely envelops the embryo with (haploid/diploid/triploid) __________________ cells and provides it with an adequate supply of stored food.

The mature angiosperm seed, like the seed of a gymnosperm, possesses three main parts: seed coats, embryo, and stored food. The difference between these two kinds of seeds is that the stored food enveloping the angiosperm embryo is the __________________, while the surrounding tissue of a gymnosperm embryo is the

__________________.

- - - - - - - - - - - - - - - - - -

triploid; endosperm; megagametophyte

43. Identify each of the lettered structures in the drawings of an angiosperm and a gymnosperm seed.

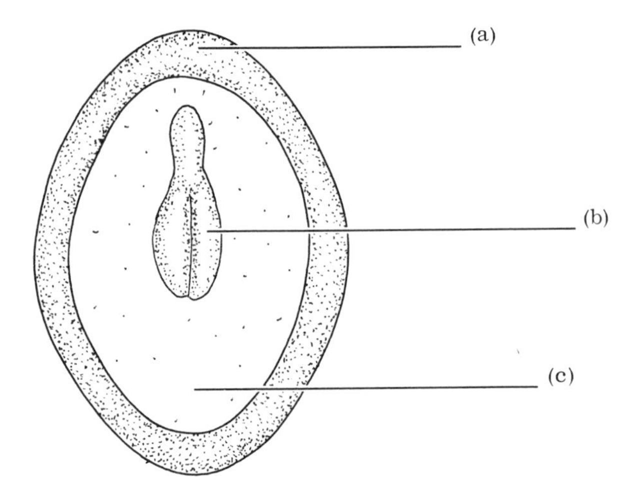

A. an angiosperm seed

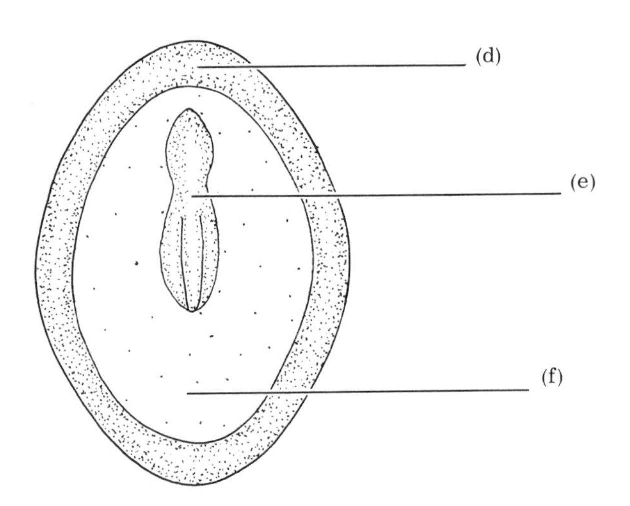

B. a gymnosperm seed

- - - - - - - - - - - - - - - - - - - -

a. seed coat; b. embryo; c. endosperm; d. seed coat; e. embryo; f. megagametophyte

44. Indicate the chromosome number—$\underline{n}$, $2\underline{n}$, or $3\underline{n}$—for each of the items

you identified in frame 43. ______________________________

____

- - - - - - - - - - - - - - - - - -

a. 2n; b. 2n; c. 3n; d. 2n; e. 2n; f. n

45. Alternation of generations has been the unifying theme in the life cycles of the plants studied in this book. You have studied alternation of generations in the life cycles of moss, Marchantia, fern, Selaginella, cycad, and pine.

How does the angiosperm life cycle fit into the scheme of alternation of generations? Study the diagram below and fill in the blanks to complete the life cycle scheme.

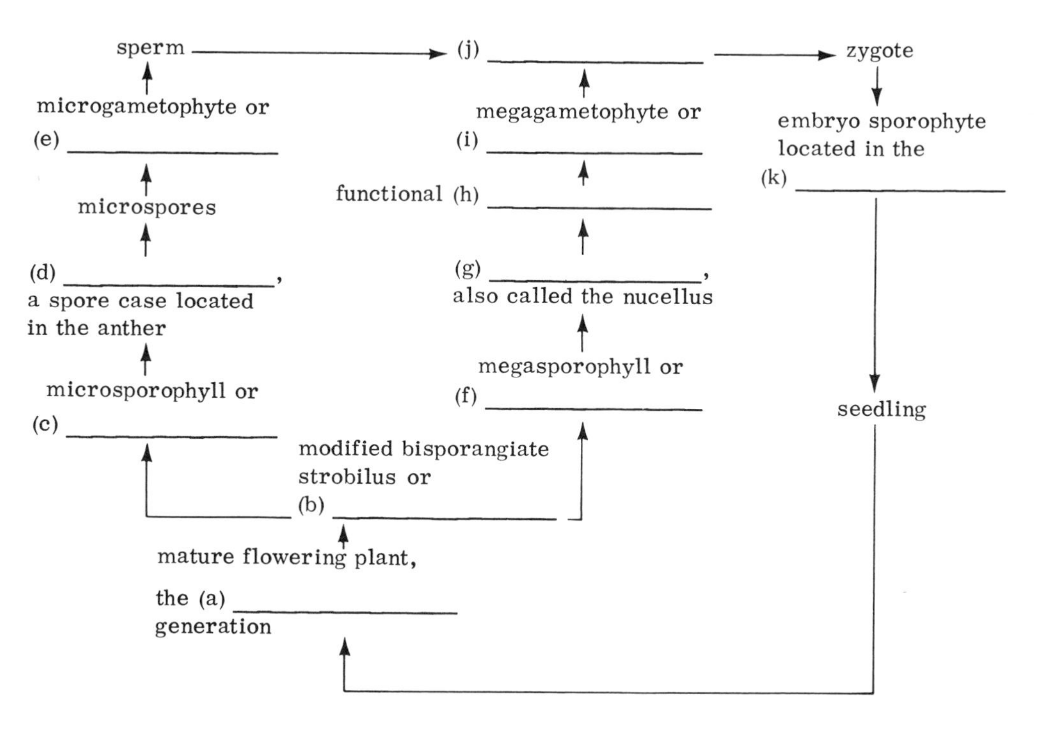

- - - - - - - - - - - - - - - - - -

a. sporophyte; b. flower; c. stamen; d. microsporangium; e. pollen grain; f. carpel; g. megasporangium; h. megaspore; i. embryo sac; j. egg; k. seed

46. Using the letters to represent the stages of the life cycle shown by the diagram in frame 45, between what two letters in the sequence f-g-h-i-j does meiosis occur in the angiosperm life cycle?

____

\- - - - - - - - - - - - - - - - - - -

between g (megasporangium) and h (functional megaspore)

47. Which of the lettered items in the diagram in frame 45 are part of the diploid sporophyte generation? ____________________ Which are part of the haploid gametophyte generation? ____________________

\- - - - - - - - - - - - - - - - - - -

sporophyte: a, b, c, d, f, g, k; gametophyte: e, h, i, j

48. Let's apply your understanding of alternation of generations in the life cycle of an angiosperm to the following problem. Suppose that the chromosome number in the cells of the petals of a tulip flower is found to be 24. How many chromosomes should be found in each of the following tulip cells?

a. polar nucleus of embryo sac _______

b. leaf cell nucleus _______

c. microspore nucleus _______

d. endosperm nucleus _______

e. egg nucleus _______

f. nucleus in a cell of the filament of a stamen _______

g. functional megaspore nucleus _______

h. fusion nucleus of embryo sac _______

i. nucleus in a cell of the embryo _______

\- - - - - - - - - - - - - - - - - - -

a. 12; b. 24; c. 12; d. 36; e. 12; f. 24; g. 12; h. 24; i. 24

SELF-TEST
Kinds of Seed Plants: Angiosperms

As a review of this chapter evaluate your understanding of angiosperms by taking this self-test. All questions should be answered in the light of information presented in Chapter 6 and preceding chapters of this book. Answers are given at the end of the test.

1. As we have noted, one problem in studying plant reproduction is the duplication of terminology where alternative terms actually refer to the same structure. Match the numbered alternatives with a lettered term.

_______ a. embryo sac
_______ b. pollen grain
_______ c. stamen
_______ d. carpel
_______ e. sepal
_______ f. flowering plant
_______ g. flower
_______ h. nucellus
_______ i. tricarpellate pistil
_______ j. ovule

1. megasporophyll
2. megasporangium
3. angiosperm
4. composed of three megasporophylls
5. microspore
6. nonessential modified leaf
7. gymnosperm
8. microsporophyll
9. megaspore
10. megagametophyte
11. a potential seed
12. modified bisporangiate strobilus
13. microgametophyte

2. The answer to each of the following questions is a number. Give the required number.

a. The angiosperm microgametophyte contains how many sperms?

____

b. The angiosperm megagametophyte contains how many eggs?

____

c. The angiosperm megagametophyte contains how many nuclei?

____

d. A bicarpellate pistil is formed from how many fused megasporophylls? _______

e. A typical flower produces how many different kinds of sporangia? _______

f. How many polar nuclei are there in an angiosperm megagametophyte? _______

g. How many functional megaspores are there in an angiosperm megasporangium? _______

h. How many nuclei give rise to the endosperm nucleus? _______

3. Label the parts of the flower in the diagram shown below.

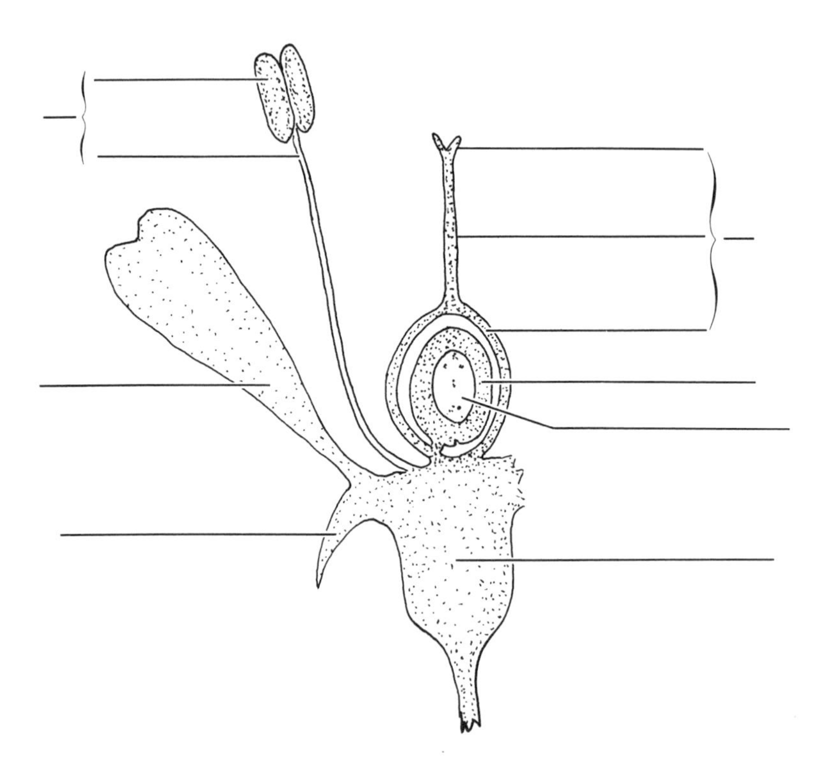

4. A typical flower consists of four whorls of modified leaves—some essential and some nonessential to reproduction. Which of the whorls of the flower shown in the diagram in question 3 are nonessential for seed production? _______________

5. This book presents the widely held view that flowering plants are descended from gymnosperms. What evolutionary sequence of steps must the megasporophyll of a hypothetical gymnosperm have undergone to produce the angiosperm pistil?

6. Trace megagametophyte development in the life cycle of a typical angiosperm by completing the following sequence.

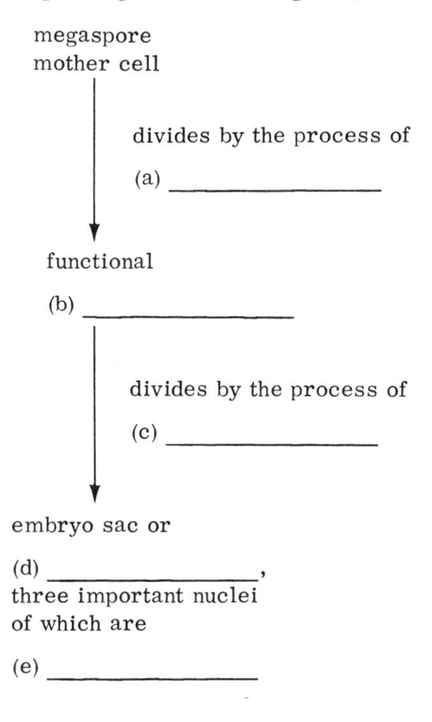

7. Trace microgametophyte development in the life cycle of a typical angiosperm by completing the following sequence.

microspore
mother cell

divides by the process of

(a) ________________

microspores

divide by the process of

(b) ________________

microgametophytes or

(c) ________________,
each of which contains two
important nuclei called

(d) ________________

8. The cells in the root of a corn plant are studied and 20 chromosomes are found in each of several cells. How many chromosomes should be found in each of the following corn cells?

   a. nucleus in cell of embryo in corn kernel _______

   b. functional megaspore nucleus _______

   c. egg nucleus _______

   d. microspore nucleus _______

   e. polar nucleus of embryo sac _______

   f. leaf cell nucleus _______

   g. endosperm nucleus _______

   h. nucleus in cell of filament of a stamen _______

   i. fusion nucleus of embryo sac _______

9. In Chapter 1 of this book we presented a brief outline of alternation of generations and you were asked to place the following stages and processes in correct chronological order to show the cyclic relationship:

   (A) spores, (B) gametes, (C) meiosis, (D) fertilization, (E) gametophyte, (F) sporophyte, (G) zygote.

Now for one final review, again put these words in correct order to represent how alternation of generations occurs in an angiosperm. Then answer questions a—f below about alternation of generations as it applies to the life cycle of an angiosperm.

a. List all of the diploid stages in the life cycle. ______________

____

b. If you have a tomato plant bearing five tomatoes growing in your garden, which generation—gametophytic or sporophytic—does this plant represent? ______________

c. Justify or explain the following: "In angiosperms there are two different gametophytes." ______________

____

d. Why is a hay-fever victim botanically correct when he says "I suffer from an allergy to ragweed microgametophytes."

____

____

e. In terms of chromosome number the sporophyte generation begins with what cell? ______________

f. Explain the statement "Angiosperms are heterosporous."

____

____

____

10. The most unique feature of angiosperms is the production of flowers. Which of the following statements about flowers and flowering plants are true?

_______ a. The flower is always a part of the sporophyte generation.

_______ b. The anther of a stamen is a place where meiosis occurs.

_______ c. <u>Both</u> sperms in the angiosperm microgametophyte unite with nuclei in the embryo sac.

_______ d. Botanists consider the flower to be a modified cone or strobilus.

Answers

1. a. 10 (frame 28)
   b. 13 (frames 13, 18)
   c. 8 (frames 11, 18)
   d. 1 (frame 19)
   e. 6 (frame 9)
   f. 3 (frame 1)
   g. 12 (frame 23)
   h. 2 (frame 18)
   i. 4 (frame 19)
   j. 11 (frame 15)

2. a. 2 (frame 33)
   b. 1 (frame 27)
   c. 8 (frame 28)
   d. 2 (frame 19)
   e. 2 (frame 23)
   f. 2 (frame 29)
   g. 1 (frame 28)
   h. 3 (frames 35, 36)

3.

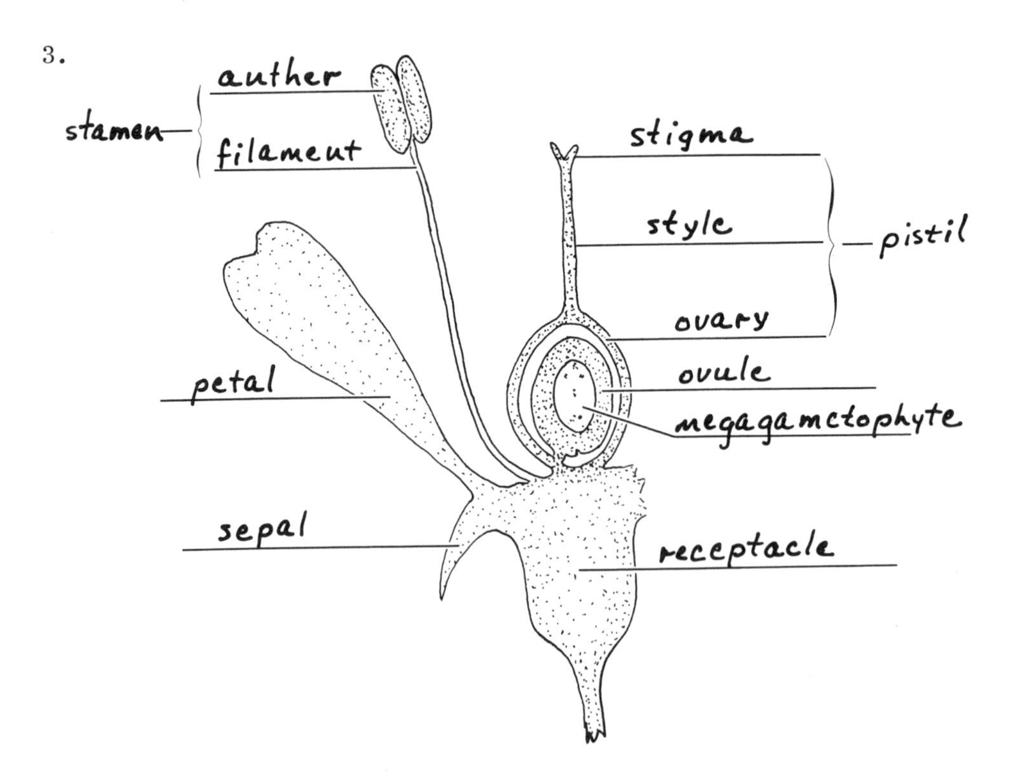

(frames 3, 5, 6, 30)

4. The sepals and petals are nonessential for seed production. (frame 9)

5. Your answer should include the following points: (1) The hypothetical gymnosperm must bear ovules on the edge of its megasporophylls. (2) Each megasporophyll must curl its edges inwardly so that the ovules become enclosed by the megasporophyll. (3) The tip of the megasporophyll becomes receptive to pollen and forms the stigma. The remaining portion of the megasporophyll becomes differentiated into style and ovary. (frames 15, 16)

6. a. meiosis; b. megaspore; c. mitosis; d. megagametophyte; e. egg and two polar nuclei (frames 27-30, 45)

7. a. meiosis; b. mitosis; c. pollen grain; d. sperms (frames 11-13, 33, 45)

8. a. 20 (frames 45-47)
   b. 10 (Chapter 5, frame 14)
   c. 10 (frame 35)
   d. 10 (frame 11)
   e. 10 (frame 35)
   f. 20 (frames 45-47)
   g. 30 (frame 37)
   h. 20 (frames 45-47)
   i. 20 (frame 35)

9.

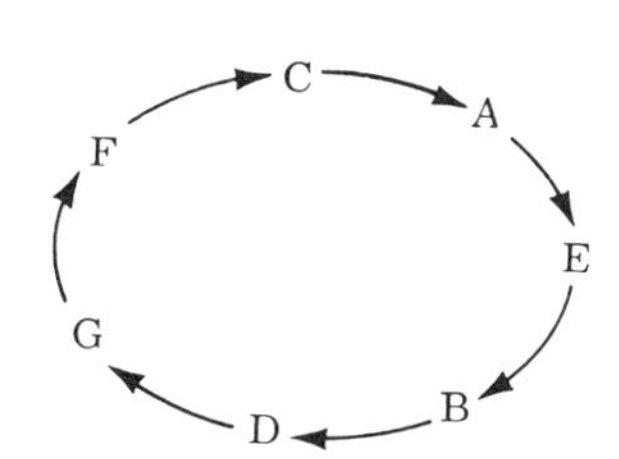

(Chapter 1, frames 56-58; Chapter 6, frames 45-47)

a. F. sporophyte and G. zygote (Chapter 1, frames 56-58; Chapter 6, frames 45-47)
b. sporophytic (Chapter 1, frames 56-58; Chapter 6, frames 45-47)
c. There are, in fact, two gametophytes in angiosperms: the microgametophyte or pollen grain which contains sperms, and the megagametophyte or embryo sac which produces an egg. (frame 45)
d. Hay fever may be due to an allergy to ragweed pollen. A pollen grain is a microgametophyte. (Chapter 5, frame 28)

e. zygote (Chapter 1, frame 57)

f. Angiosperms produce microspores in the anthers of the stamens and a functional megaspore in each ovule in the ovary of the pistil. Since they produce two different kinds of spores, they are said to be heterosporous. (frame 45)

10. All four statements are true. (a. frames 23, 45; b. frame 11; c. frames 36–38; d. frame 23)

FINAL TEST

1. Listed below are a number of stages and processes that occur in alternation of generations. Put these lettered items in correct chronological order by letter, beginning with item (F) sporophyte.

   (A) gametophyte, (B) spores, (C) zygote, (D) meiosis, (E) fertilization, (F) sporophyte, (G) gametes.

   ____

2. The following items are parts of a Selaginella plant. Arrange them from the largest and most complex to the simplest and smallest.

   (A) microsporangia, (B) strobilus, (C) microspores, (D) sporophyte, (E) microsporophylls.

   ____

3. To summarize the life cycle of a pine, study the diagram below and fill in the names of the missing structures.

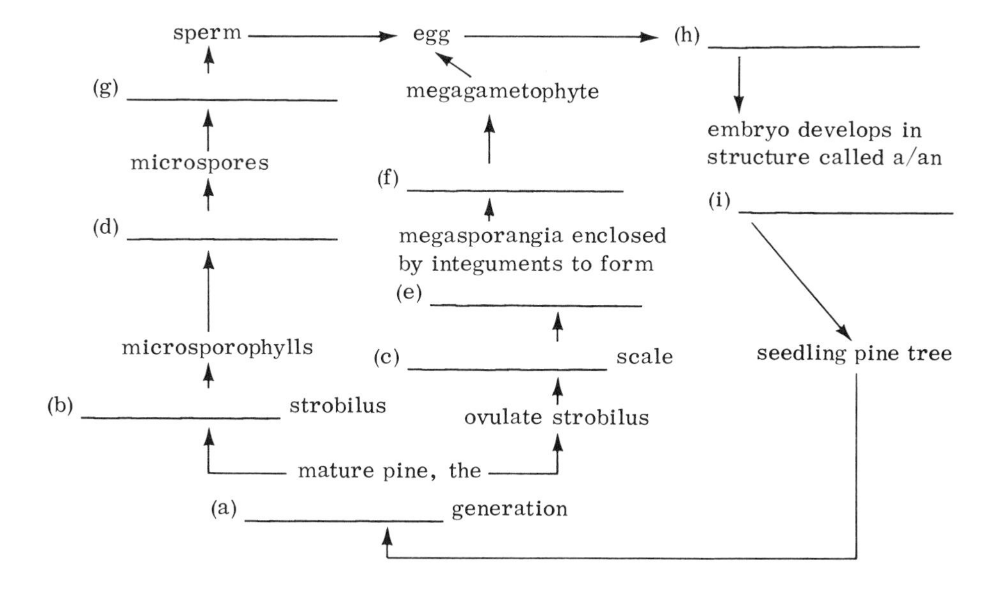

4. Identify the lettered items in the diagram of the moss plant on the next page. Indicate in the parentheses whether each structure is n or 2n.

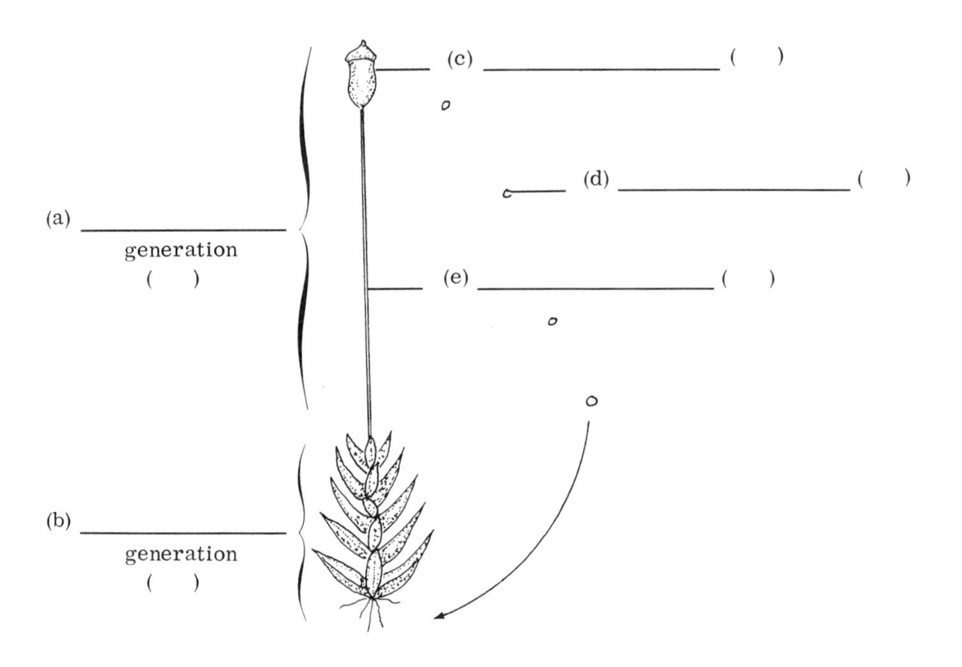

5. Shown below is a drawing of part of a heterosporous plant. Label the lettered items in the drawing.

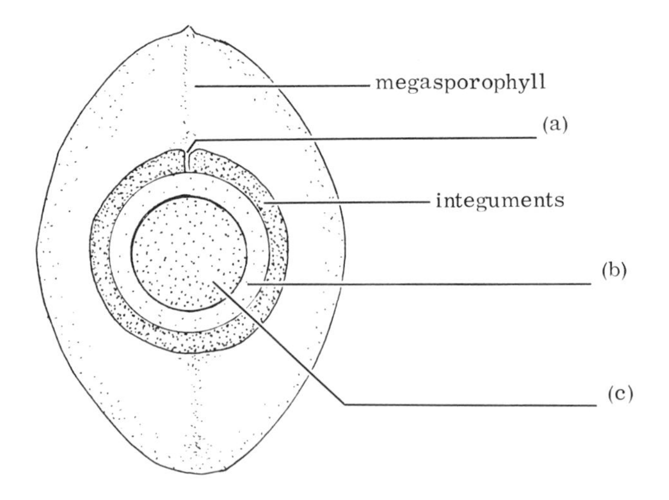

6. If we symbolize mitosis as 2n ⟶ 2n and meiosis as 2n ⟶ n, then we may symbolize fertilization as

_______ a. $3\underline{n} \longrightarrow 3\underline{n}$

_______ b. $\underline{n} + \underline{n} = 2\underline{n}$

_______ c. $2\underline{n} \longrightarrow 4\underline{n} \longrightarrow 2\underline{n}$

_______ d. $\underline{n} \longrightarrow \underline{n}$

_______ e. $\underline{n} \longrightarrow 3\underline{n}$

7. In sexually reproducing organisms reproductive cells unite to initiate a new individual. These reproductive cells may be both morphologically and physiologically different or they may differ only physiologically. Which of the following is a correct statement about the facts just described?

_______ a. When sex cells can be differentiated both morphologically and physiologically, they are spoken of as isogametes.

_______ b. Isogametes appear to be the same under the microscope but they are physiologically different.

_______ c. Selaginella is an example of an isogamous plant.

_______ d. Heterogametes appear to be different under the microscope but they are physiologically the same.

8. This book has emphasized the evolutionary progression of plants from relatively primitive to more advanced forms. From the following pairs of items select the more advanced member of each pair and write its number in the blank.

_______ a. (1) Plants with relatively large, conspicuous gametophyte; (2) plants with large, conspicuous sporophyte.

_______ b. (1) Plant with heterospory; (2) plant with homospory.

_______ c. (1) Plant with swimming sperms; (2) plant with sperms lacking structures for locomotion.

_______ d. (1) Vascular tissue lacking in sporophyte; (2) vascular tissue present in sporophyte.

_______ e. (1) Plants retaining megaspores in the megasporangium; (2) plants releasing the megaspores from the megasporangium.

_______ f. (1) Cone-bearing plants (conifers); (2) flowering plants.

9. Check the true statements about terrestrial plants among the following.

_______ a. The gametophyte is less conspicuous than the sporophyte.

_______ b. Vascular tissue is limited to the sporophyte generation.

_______ c. Sporophylls never occur in the gametophyte generation.

_______ d. Spores germinate to form gametophytes.

_______ e. All vascular plants produce seeds.

10. Indicate which of the following statements are true and which are false.

_______ a. Pollination always precedes fertilization in the seed plants.

_______ b. Seed plants are always heterosporous.

_______ c. In seed plants the megaspore is retained and develops into a megagametophyte within the megasporangium.

_______ d. A seed may be defined as an embryo surrounded by megasporangium and integuments.

_______ e. Although water must be present for sperms to reach the egg in a moss plant, water is not necessary for fertilization in seed plants.

11. Over the years that sexual reproduction of seed plants has been studied some terminology has been duplicated. In the following exercise match the numbered items with their equivalent lettered item.

_______ a. Pollen grain = _____.

_______ b. Megasporangium plus integuments = _____.

_______ c. Megasporangium is the same as _____.

_______ d. Female gametophyte = _____.

_______ e. Transfer of the male gametophyte to the vicinity of the female gametophyte is called _____.

1. egg
2. fertilization
3. ovule
4. integuments
5. microgametophyte
6. pollination
7. nucellus
8. microsporophyll
9. microspore
10. megagametophyte
11. seed

_______ f. A/an _____ is an ovule containing an embryo.

_______ g. The hard, dry, protective seed coats are derived from the _____.

12. Match numbered items with the most appropriate lettered items. Use numbered items more than once if necessary.

_______ a. microgametophyte

_______ b. homosporous plant

_______ c. structure containing a single megagametophyte

_______ d. constituents of a _Selaginella_ strobilus

_______ e. produced by meiosis in _Selaginella_

1. _Selaginella_
2. megasporophylls
3. megaspore
4. female gametophyte
5. liverwort
6. sperms
7. male gametophyte
8. microsporangium
9. microsporophylls
10. produces eggs
11. large gametophyte

13. In _Selaginella_ microspores are produced in spore cases called ____________________, located on modified leaves called ____________________, which constitute a part of a conelike structure called a/an ____________________.

14. A green leafy portion of a seed plant constitutes the (gametophyte/sporophyte) ____________________ generation and as such produces (gametes/spores) ____________________ by the process of meiosis. The meiotic products are produced in structures called ____________ which are borne on modified leaves called ____________________.

15. An angiosperm embryo is immediately surrounded by the nutritive tissue called ____________________ which has the ________________ chromosome number, while a gymnosperm embryo is immediately surrounded by the nutritive tissue called the ____________________ which has the ____________________ chromosome number.

16. Indicate which of the following structures are sporophytic and which are gametophytic.

    a. sporophyll ___________________

    b. vascular tissue ___________________

    c. embryo sac ___________________

    d. fern leaf ___________________

    e. moss leaf ___________________

    f. pollen grain ___________________

    g. Marchantia capsule ___________________

17. Listed below are statements which describe conditions existing in plants reproducing solely by sexual or solely by asexual means. Write "sexual" or "asexual" after each statement to indicate the method of reproduction.

    a. This plant species would probably be adapted to a wide variety of environments. ___________________

    b. Meiosis occurs in the course of reproduction in this plant.

       ____

    c. All of the offspring of a single specimen of this plant species would be genetically identical. ___________________

    d. An alternation of generations may exist in the life cycle of this plant. ___________________

18. If you found that a certain watermelon contained 436 mature seeds, at least how many

    a. ovules were originally present? _______

    b. pollen tubes had to function in producing the 436 mature seeds?

       ____

    c. functional sperms were formed in producing the 436 mature seeds? _______

    d. functional megaspores produced viable megagametophytes?

       ____

19. Indicate what each of the following items becomes or develops into:

a. megaspore mother cell ______________________

b. ovule ______________________

c. microspore ______________________

d. integuments ______________________

e. functional megaspore ______________________

f. zygote ______________________

g. two polar nuclei + a sperm nucleus ______________________

h. ovuliferous scales in pine cone ______________________

20. The common sunflower has a diploid chromosome number of 34. How many chromosomes should be present in the following sunflower cell nuclei?

a. egg _______

b. megaspore mother cell _______

c. microspore _______

d. endosperm _______

e. nucellus _______

f. polar nucleus of embryo sac _______

g. leaf cell _______

h. stigma of pistil _______

21. All of the following are parts of or stages in the Selaginella life cycle: (A) microspore, (B) megasporophyll, (C) megagametophyte, (D) zygote, (E) strobilus, (F) sperm, (G) green leafy club moss, (H) microsporophyll, (I) microgametophyte, (J) sporophyte. Which of these ten stages are part of the gametophyte generation?

____

22. Suppose you found a certain simple plant species which reproduced only by asexual means. Could alternation of generations be demonstrated in this plant species? __________ Explain your answer.

____

____

23. What is the structural relationship between sporophylls, strobili, and sporangia? ______________________________

____

24. Upon encountering a seed plant unknown to you, in what four ways can you expect this plant to differ from a non-seed plant?

____

____

____

____

25. What evolutionary advantage is gained by a plant that produces seeds?

____

____

Answers

1. (F) sporophyte, (D) meiosis, (B) spores, (A) gametophyte, (G) gametes, (E) fertilization, (C) zygote (Chapter 1, frames 55-57)

2. (D) sporophyte, (B) strobilus, (E) microsporophylls, (A) microsporangia, (C) microspores (Chapter 3, frame 24)

3. a. sporophyte; b. staminate; c. ovuliferous; d. microsporangia; e. ovule; f. megaspores; g. microgametophyte; h. zygote; i. seed (Chapter 5, frames 24-34)

4.

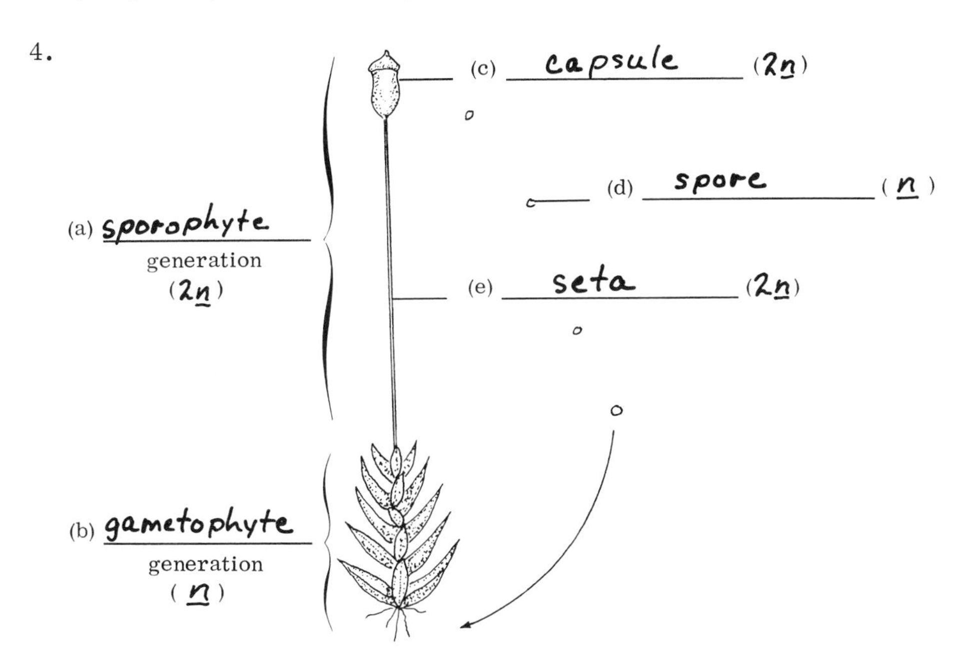

(Chapter 2, frames 13, 14)

5.

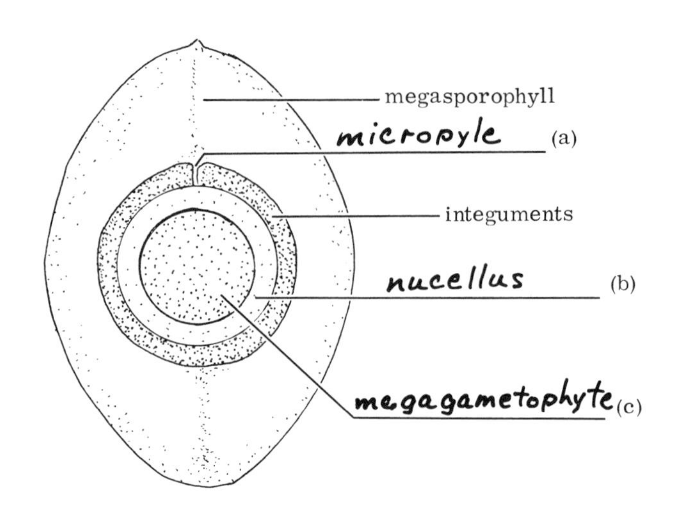

(Chapter 4, frame 24)

6. b. n + n = 2n (Chapter 1, Self-Test question 4)

7. b is correct (Chapter 1, frames 16-25; Chapter 3, frame 24)

8. a. 2 (Chapter 2, frames 24-26)
   b. 1 (Chapter 3, frames 4, 5; Chapter 4, frame 12)
   c. 2 (Chapter 2, frame 40; Chapter 5, frame 29)
   d. 2 (Chapter 2, frame 27)
   e. 1 (Chapter 4, frame 12)
   f. 2 (Chapter 6, frame 2)

9. a. true (Chapter 2, frame 26)
   b. true (Chapter 2, frame 27)
   c. true (Chapter 2, frame 31)
   d. true (Chapter 1, frame 54)
   e. false (Chapter 4, frame 5)

10. All five statements are true. (Chapter 4, frames 11, 12, 30, 33, 35)

11. a. 5 (Chapter 4, frame 28)
    b. 3 (Chapter 4, frame 20)
    c. 7 (Chapter 4, frame 21)
    d. 10 (Chapter 3, frame 14)
    e. 6 (Chapter 4, frame 30)
    f. 11 (Chapter 4, frame 35)
    g. 4 (Chapter 4, frame 37)

12. a. 7 (Chapter 3, frame 14)
    b. 5 (Chapter 2, frame 19; Chapter 3, frame 4)
    c. 3 (Chapter 4, frame 14)
    d. 2 and 9 (Chapter 3, frame 18)
    e. 3 (Chapter 3, frame 20)

13. microsporangia, microsporophylls; strobilus (Chapter 3, frames 18, 19)

14. sporophyte; spores; sporangia; sporophylls (Chapter 4, frames 2, 11, 16; Chapter 5, frames 22, 34; Chapter 6, frame 45)

15. endosperm; triploid; megagametophyte; haploid (Chapter 6, frame 42)

16. a. sporophytic (Chapter 2, frames 30, 31)
    b. sporophytic (Chapter 2, frame 27)
    c. gametophytic (Chapter 6, frame 28)
    d. sporophytic (Chapter 2, frame 28)
    e. gametophytic (Chapter 2, frame 3)
    f. gametophytic (Chapter 4, frame 28)
    g. sporophytic (Chapter 2, frame 19)

17. a. sexual; b. sexual; c. asexual; d. sexual (Chapter 1, frames 6, 13, 34, 56)

18. a. 436; b. 436; c. 872; d. 436 (Chapter 4, frames 13, 32, 35)

19. a. four megaspores (Chapter 5, frame 14)
    b. seed (Chapter 4, frame 35)
    c. microgametophyte or pollen grain (Chapter 3, frame 12)
    d. seed coats (Chapter 4, frame 37)
    e. megagametophyte (Chapter 5, frame 15)
    f. embryo sporophyte (Chapter 1, frame 56; Chapter 4, frame 2)
    g. endosperm nucleus (Chapter 6, frames 35, 36)
    h. winglike extensions on pine seeds (Chapter 5, frame 32)

20. a. 17 (Chapter 6, frame 35)
    b. 34 (Chapter 5, frame 14)
    c. 17 (Chapter 6, frame 11)
    d. 51 (Chapter 6, frame 37)
    e. 34 (Chapter 4, frame 21)
    f. 17 (Chapter 6, frame 35)
    g. 34 (Chapter 6, frames 45-47)
    h. 34 (Chapter 6, frames 15, 16)

21. (A) microspore, (C) megagametophyte, (F) sperm, and (I) microgametophyte (Chapter 3, frame 24)

22. No. Alternation of generations occurs only in sexually reproducing species of plants where a haploid gametophyte alternates with a diploid sporophyte. (Chapter 1, frames 52-56)

23. Sporangia are located on sporophylls which are clustered in conelike structures called strobili. (Chapter 3, frame 16)

24. A seed plant (1) is heterosporous; (2) possesses one functional megaspore in each megasporangium; (3) retains the single functional megaspore within the megasporangium; and (4) possesses integuments enclosing the megasporangium. (Chapter 4, frame 18)

25. Since the seed protects its enclosed embryo from desiccation and freezing, the seed increases the chances of survival for the plant species. (Chapter 4, frame 5)

Index

Pages indicating introductions, summaries, and self-tests are not included in this index.